371109

302.2

301.H6 MAN

KU-511-211

New Communication Environments

5295084 8MM

"Giuseppe Mantovani has provided contemporary scholars with a conceptually rich and compelling way to think about human thought and action in the new communication environments that pervade many spheres of modern life. He does a brilliant job of bringing together scholarship from an astonishing range of disciplines in a manner that demands and rewards repeated reading and reflection. I highly recommend this book to anyone interested in the intersecting concerns of cognitive science, communication studies, organization theory, and social theory."

Michael Cole, University of California at San Diego

"A thorough and wide-ranging discussion of recent debates within the cognitive and social sciences and their relevance to our understanding of computer systems, cooperative work and virtual environments. An informative and timely book which will prove of interest to those from various disciplines who are concerned with technological innovation and interpersonal communication."

Christian Heath, King's College, University of London

"This is an important work in pressing the boundaries of cognitive theory so as to comprehend the contemporary communications maelstrom. In creatively expanding the sensitivity to issues of shared meaning and identity, Mantovani gives us critical tools for dealing with new communication environments."

Kenneth J. Gergen, Swarthmore College

"Overall, I believe that this book is an important review of sound social studies of computerization, which is accessible to non-specialists and which should be read by anybody who is interested in conceptualizing new electronic media, including social scientists, computer scientists, and information scientists."

Rob Kling, University of California at Irvine

New Communication Environments

From Everyday to Virtual

GIUSEPPE MANTOVANI

Taylor & Francis
Publishers since 1798

UK   Taylor & Francis Ltd, 1 Gunpowder Square, London EC4 3DE
USA   Taylor & Francis Inc., 1900 Frost Road, Suite 101, Bristol PA 19007

Copyright © Giuseppe Mantovani 1996

All rights reserved. No part of this publication may be reproduced, stored in a retrieval system, or transmitted, in any form or by any means, electronic, electrostatic, magnetic tape, mechanical, photocopying, recording or otherwise, without the prior permission of the copyright owner.

British Library Cataloguing in Publication Data

A catalogue record for this book is available from the British Library

ISBN 0-7484-0395-7 (cloth)
ISBN 0-7484-0396-5 (paperback)

Library of Congress Cataloging in Publication Data are available

Cover design by Hybert Design & Type

Typeset by Mathematical Composition Setters Ltd,
Salisbury, Wiltshire, SP3 4UF

Printed in Great Britain by T. J. Press (Padstow) Ltd, Padstow.

SOUTHAMPTON INSTITUTE
LIBRARY SERVICES LTD

SUPPLIER   25

ORDER No

DATE   2.6.99

Contents

9 Conclusions: Electronic Communication, Social Context and Virtual Identities

Preface

Owing to its content and style, this book is not only for students in social psychology, social and cognitive sciences, and communications, but also for those carrying out research or working in such areas as media development and use, corporate communications, organizational studies, educational technologies and tele-didactics. Professionals in communication technologies may also be interested, as a considerable amount of their work deals with interactions between social actors and technological artifacts in everyday working or living conditions. Attention has been paid to rendering the problems treated clear and essential, without descending into excessive detail of the points on which extensive scientific literature exists (e.g., social identity, processes of construction of the self, etc.), so as to make this volume accessible to readers without previous specific knowledge of cognitive and social sciences.

G. MANTOVANI

Acknowledgements

First and foremost, my debt to eminent scientists like Michael Cole, James March, Donald Norman, Marshall Sahlins, Karl Schmidt and Lucy Suchman, who suggested many of the ideas developed in this text, is gratefully acknowledged and is duly emphasized by the many quotations from their works to be found in these pages.

The studies of social psychologists like Martin Lea, Janet Fulk and Russell Spears helped me substantially to build my three-level model of social context: to them I owe the basic idea that social reality is something deeper and richer than mere interpersonal relationships. Dominic Abrams and John Turner were no less important to me: they elucidated processes like self-categorization and social identity formation, which link individual conduct to the normative dimension of social contexts.

I am also deeply indebted to Philip Agre, Bill Clancey, Tom Hewett, Rob Kling and Ray Loveridge for their encouragement and interest in the perspectives expressed here.

I am very grateful to Paolo Legrenzi, and also to Bruno Bara, Antonella Carassa, Erminio Gius, Arrigo Pedon and Anne Maass, who were among the many friends and colleagues who gave me their valuable advice, attention and time. I would also like to thank pre- and postgraduate students and young research workers at the University of Padova, whose interest in the topics considered here I found highly stimulating.

I am especially grateful to Gabriel Walton, a good friend of mine, who made this translation from the original Italian. The quality of her faithfully rendered work will be apparent to all readers. I had the privilege of appreciating by personal experience her subtlety, sparkle and sense of humour, all of which made our collaboration a pleasure.

Last, but not least, are my debts to Richard Steele, publisher, who encouraged the publication of this book, and to Taylor & Francis, Ltd, for having included it in this prestigious series.

The conceptual model of social context presented in Chapter 4 is illustrated and discussed in a paper in *Cognitive Science* (1996) addressing the generation of high-level mental models of interactive systems. Chapter 7, on computer-mediated communication, expands ideas which were previously compressed in an article in *Human Relations* (1994). Chapter 8 on virtual reality as a communication environment is mainly based on the text of a paper which appeared in *Human Relations* (1995).

G. MANTOVANI

Introduction:

From Uses to Social Actors in Context

We are now entering a novel phase in the history of human–computer relationships, a phase in which some once dominant issues will become quite obsolete, while previously unasked questions will arise. A few years ago, the most revered among cognitive artifacts were expert systems, which many thought were even destined to supplant human decision makers in a number of complex diagnostic activities. Our attention was then riveted on key words like *intelligent systems*, *interfaces*, *menus*, *designers*, *responsibility* and *control* over systems by users. Now emphasis is shifting towards areas like *computer-mediated communication* (CMC), *computer-supported cooperative work* (CSCW), and *virtual reality* (VR) as media. We can still worry about *users* and consider their ways of interacting with computers as matters worthy of our interest, but priorities in both practice and research are changing fast. We no longer have to face and model system users but social *actors*, who are participating in electronic environments in order to reach their peculiar goals, to assert their principles and values, and to develop their projects and self-identities.

This book considers three main themes: the cognitive perspective of *situated action* (Part 1), a conceptual *model of contexts* as normative social structures affecting artifact use (Part 2), and the individual and organizational impact of the new *electronic environments of cooperation and communication* (Part 3). Connections between the above themes are important. From Part 1 to Part 3, we see that the situated action approach provides us with a firmly based socio-cognitive scenario fitting our conception of both contexts and artifacts as incorporating normative social structures, which in turn gives us access to the social processes involved in the current diffusion of information technology tools. The link can also be traced the other way around, from Part 3 to Part 1: communication environments convey cultural models and social norms which in turn inspire purposeful situated actions aiming at projecting socially recognizable self-identities. *Communication* and *identity* are in fact the two poles of our discussion, which

1

explores how culture and technology together shape the situations we live in and influence the development of our social and individual identities.

Part 3 of this volume is devoted to an analysis of the *environments constructed by new communication technologies*. The changes that have taken place in the last ten years reveal how rapidly technologies have both progressed and regressed in such a short space of time. Their advance is apparent in their increasing capacity to permeate the everyday lives of all of us, even though we may not be professionals working on CMC. Today, computer systems can act more indirectly than ever before, silently pervading significant areas of our daily lives, so that in many cases there is little difference between users and non-users. In a certain sense regression also occurs, since technologies in communication environments often do not openly occupy the stage but seem to vanish into the background. This is to be expected, because as soon as a technology reaches the degree of maturity and reliability that really makes it usable, attention switches from the technological aspects of artifacts – which were previously difficult to master and therefore of great concern for non-specialist users – to the activities, projects and goals of social actors using artifacts for their own purposes.

The theoretical scene has also changed. A new socio-cognitive approach, that of *situated action*, has been developed, which seems to us to explain some of the issues of the new communication environments. Part 1 of this book discusses situated action and elucidates its significance with simple and, we hope, pertinent examples drawn from everyday experience and from the works of authors ranging from Jane Austen to John Fowles. The situated action perspective, as will be made clear in the following pages, is not an alternative to the traditional symbolic approach to human information processing. It only shifts emphasis from cognition to action, from decision-making as a sort of calculus to intuitive decision as an evaluation made in terms of appropriateness and self-image, from individual cognitive processes to social action in its normative and cultural context.

From the situated action point of view, action is no longer regarded as the simple execution of pre-established plans, but as a flexible adaptation to the peculiarities of circumstances. This perspective leads us to see that 'the contingence of action on a complex world of objects, artifacts and other actors, located in space and time, is no longer treated as an extraneous problem with which the individual actor must contend, but rather is seen as the essential resource that makes knowledge possible and gives action its sense' (Suchman, 1987, p.179). Situated action poses many questions, all of great moment for the cognitive and social sciences, such as the ambiguity of situations, possible inconsistencies in actors' interests, the development of opportunistic strategies, and the role of principles and values in the construction of self-images.

In addressing these questions, we rely mainly on two frameworks, which we consider of particular pertinence and high theoretical profile. On the one hand is the approach to decision-making known as the *interpretation of situations*, as developed by James March; on the other is the idea that intuitive decision-making is a matter *more of appropriateness than of calculus*, as upheld by Lee Beach and Terence Mitchell. We think that these two reference points may make an important contribution to our understanding of actors' conduct in daily situations.

Between the presentation of the situated action approach and the analysis of the social impact of electronic environments stands the second, central, part of this book, which acts as hinge between the themes treated in Parts 1 and 3. Part 2 is the very core of our discourse. It has two foci: one is a *three-level conceptual model of social context* considered as symbolic order which actors receive and regenerate constantly in action; the other develops a framework for *understanding artifacts*, both technological and conceptual, as tools embodying socially recognizable values and practices.

These two issues are complementary: the first traces a top-down path, from cultural order to artifacts in daily use; the second completes the circle through a bottom-up route ascending from local practices to meanings ascribed to situations by cultural order. Our conceptual model of social context draws partly on Donald Norman's views regarding cognitive artifacts, and partly on Michael Cole's cultural psychology, but above all it profits from Marshall Sahlins' ideas about cultural systems supplying actors with sets of shared interpretative criteria providing situations with appropriate meanings.

That such a hinge is useful is confirmed by the fact that research on computer-mediated communication, computer-supported cooperative work and distributed artificial intelligence appears to be in deep waters when it tries to explain how social actors, although starting from different experiences and unreconciled visions of the world, are still able to understand each other within electronic communities. We believe that our model of social context can prove valuable not only for modifying the current bottom-up research agenda in human–computer interaction, but also for stimulating empirical analyses of both interactive systems and communication environments (Mantovani, 1995, 1996).

All conclusions are postponed to the end of the text, where we discuss whether the perspective of situated action (Part 1) and the model of context (Part 2) can truly help us to understand the new environments of computer-mediated communication and cooperation (Part 3).

PART ONE

The 'Situated Action' Approach

CHAPTER ONE

The New 'Black Box'

Everyday Situations

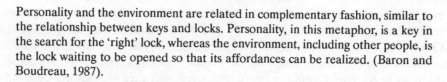

Personality and the environment are related in complementary fashion, similar to the relationship between keys and locks. Personality, in this metaphor, is a key in the search for the 'right' lock, whereas the environment, including other people, is the lock waiting to be opened so that its affordances can be realized. (Baron and Boudreau, 1987).

My point is that the metaphor system of Inside and Outside finds itself unable to make any stable sense of concepts that reside neither in the agent nor in the world, but in the relationship and interaction between the two. (Agre, 1993)

1.1   INTERESTS IN ACTORS MEET OPPORTUNITIES IN THE ENVIRONMENT

Daily life situations are currently at the centre of attention in psychological research. Social (Ross and Nisbett, 1991), environmental (Gibson, 1986), cognitive (Bara, 1995; Greeno, 1989; Lave, 1988, 1992; Resnik, 1994) and developmental psychology (Bronfenbrenner, 1979; Rogoff et al., 1994) all reflect a thriving 'ecological' sensitivity aiming at maintaining the object of investigation inside its 'natural' context.

The perspective focusing on everyday experience, which Greeno and Moore (1993) call 'situativity theory', is increasingly influencing experimental research, inspiring greater caution than in the past towards simplifications and generalizations. It also has important theoretical significance; this is manifest, for example, in the debate opened by Vera and Simon (1993) to which *Cognitive Science* recently devoted an entire issue.

However, it is not always clear why situations have characteristics which are difficult to treat by means of the theories and scientific methodologies of the positivist school, which make a too clear-cut division between subject and environment, between observer and object observed, between actor and social and physical context (Duranti and Goodman, 1992). We therefore require a model that allows us to capture that particular type of complexity concealed in the apparent simplicity of everyday interactions between people and their environments, without losing contact with the needs of scientific discourse but, on the contrary, deepening it. We propose (Chapter 4) a conceptual three-level model of the social context which expands and organizes the considerations made here with reference to situations. It extends them upwards to include symbolic order, understood as the system of meanings of a given culture, which in daily situations is continually realized and at the same time subverted by action. It also extends them downwards to contain local interactions which actors establish with their environment by means of artifacts.

This chapter deals with the second of the three levels of our model of social context: the interaction of actors with daily situations. The term 'actors', or 'social actors', designates subjects who move on their own initiative in the social and physical environment pursuing autonomously defined interests and goals. People, groups and organizations are usually called 'actors' both in social sciences (Ahrne, 1990; Alexander, 1988; Archer, 1988) and in *Distributed Artificial Intelligence* (Gasser, 1991), to stress the fact that they are able to act strategically and to exploit situations in order to satisfy their needs. Technologies and artifacts are not actors, although in *Human–Computer Interaction* some software systems are currently called 'agents' to emphasize their relative autonomy (Genesereth and Ketchpel, 1994; Maes, 1994; Minski and Riecken, 1994).

In a certain sense, the metaphor of people acting out a predetermined scenario is not really completely satisfactory. This is because situations do not exist prior to actors but are constructed by them, according to their competences and goals, both cognitively and practically. The actions of people, groups and organizations are essential components of the situation. 'The very idea of a situation means that we are not standing outside it and hence are unable to have any objective knowledge of it. We are always within the situation and to throw light on it is a task that is never entirely completed' (Gadamer, 1975, quoted in Winograd and Flores, 1986, p.29).

Analysis of psychological processes as situated operations (Lave, 1992; Resnik, 1994) eludes methods based on any kind of rigid separation between observer and object observed. It is not a coincidence that some of the most illuminating studies on the effects of new information technologies and communications on social contexts come from ethnographic research (Suchman, 1987; Zuboff, 1988), which arose precisely because of the refusal to counterpose subject and object, data and meanings.

We may initially imagine a situation as a repertory of possible information, an immense database containing a wealth of documentation on a series of problems, none of which, however, is important until all the actors, with their projects, appear on the stage. It is at this moment that the scene becomes animated, the situation is experienced, and even configurations of apparently irrelevant data are revealed as interesting. The situation becomes filled with precious opportunities and fearful threats as actors' goals, interests, strategies and interactions gradually emerge. 'It makes no sense to create a "taxonomy of situations" independent of the psychological mechanisms within humans. Psychological mechanisms evolved because they receive, process, and respond only to certain forms of environmental input' (Buss, 1991, p.481).

The consequence is that 'the dimensions of contextual input important for persons depend on the proximate goals toward which humans direct action and the specific psychological mechanism activated by each proximate goal' (*ibid.*). This is why one good metaphor of the relationship between actor and situation is that of the key and the lock. Each, alone, does not work; they must be combined properly. Each lock requires its own particular key, and vice versa. 'Personality and the environment are related in complementary fashion, similar to the relationship between keys and locks. Personality, in this metaphor, is a key in the search for the "right" lock, whereas the environment, including other people, is the lock waiting to be opened so that its affordances can be realized' (Baron and Boudreau, 1987, p.1227).

The theme of *affordance* is central to Gibson's (1986) environmental psychology. Affordance designates the instantaneous recognition of something that may satisfy an individual's needs: it may signal that a certain female is receptive, a certain tree gives shade, a field has good grazing, or a nipple gives milk. Clearly, these perceptions are formed on adaptive bases. But in the interplay between actors and environment, there are not only phylogenetically adapted components which create immediate recognition in the sense of Gibson's affordance. Cultural elements too mediate recognition of opportunities available in the environment by activating explicit or implicit, conscious or automatic processes of interpretation. In this second sense, i.e. the presence of interpretation in the relation between actors and situations, reference to affordance ceases to be completely appropriate.

We wish to stress that it is not only the environment which is constructed by actors; they too, in turn, are 'modelled' by the opportunities which that environment offers them. There is no hierarchy between key and lock, no fixed order: actors do not control situations, nor do situations control actors. Actors intervene in situations, activated and oriented by aspects of them which in turn are relevant to the extent to which actors perceive them and respond to them.

Situations change not only from one actor to another, seeing that each has her or his own particular goals, but also with respect to the same actor, because her or his system of goals is intrinsically unstable and s/he relates differently to the same situation at different moments. As actors change and their goals are transformed, so too are both the structure of the situation and the shape of the actor–situation interaction.

1.2   CONSTRUCTION OF SITUATIONS AND AVAILABLE COMPETENCES

The complexity of situations in daily experience is not to be conceived of as deriving from an excessive quantity of stimuli offered by the environment, surpassing human capacities to process information. It is true that situations are virtually inexhaustible mines and endless mazes of possible information, but we see that human beings can quite easily manage them in the course of their usual activities without too much difficulty, except in special cases, like those of certain work situations.

Actors cut the potentially available information down to size, so to speak, so that the situations they have to face are within their capabilities, and can be managed in order to meet their interests. This result is obtained by filtering both input and output, i.e. by choosing particular kinds of information to be processed and particular courses of action in response to the relevant aspects of the situation (Figure 1).

Actors simplify situations by opening variously sized windows through which they can focus on what interests them and intervene on the aspects of the situation nearest to their goals. Although the creation of these windows drastically reduces the amount of information to be examined with respect to that potentially available in the environment, it does allow deeper analysis of important aspects, thanks to the intensive allocation of attentional and practical resources.

Windows allow selections which relegate some situational aspects to the background, while they highlight others that are considered more important. Creating windows implicitly involves ordering in the priority of actors' interests: for example, why do I prefer to watch Laura instead of David during class? And why, when I pick up the newspaper in the evening, do I read the political news instead of the sports reports? In monopolizing and orienting attention, interests establish a reciprocal hierarchy and find an order, precarious though it may be.

While we cut windows in daily situations, we also interpret what we are cutting out. In fact, we cut out precisely what may be interpreted in a certain way, what

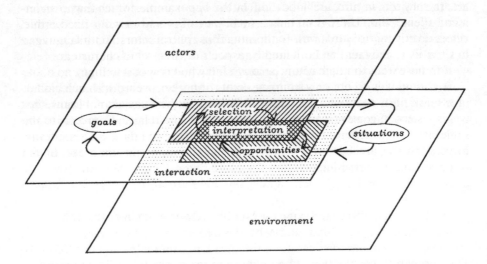

Figure 1 Actors' goals and characteristics of situations interact in a space shared by both actors and the environment.

may have a certain meaning for us. We actively construct situations: social cognition (Wyer and Srull, 1983; Zalesny and Ford, 1990; Ross and Nisbett, 1991) shows that the environment of daily experience, not only social but also physical, is not a given fact which we simply acknowledge, but is the product of social processes of categorization, stereotyping, identification with groups, formation, and transmission of values and social norms.

Social influence processes, both informational and normative, establish what is appreciable and what is not. These processes permeate the construction of our identities, the definition of our goals, and the formulation of our projects. All three are the starting points of those operations of selection and interpretation by means of which we structure situations and give them precise meaning. The starting point composed of projects, goals and identities is in turn constantly modified by the various configurations assumed by situations as they respond to our initiatives. As our goals supply the criteria to select and interpret situational aspects, so the opportunities or threats identified in situations – together with the responses that the environment makes to our initiatives – supply feedback on our system of goals, encouraging some and discouraging others.

The processes which intervene in the selection of important situational aspects and the resulting determination of intepretations are identified by the theory of stress as a cognitive–affective assessment of the appropriateness of the transaction between a particular actor and a particular aspect of the environment (Lazarus, 1991). This theme is linked to that of actors coping with opportunities or dangers perceived in daily situations; studies on stress and coping show how people's responses to situations are highly differentiated.

One first factor which we encounter when constructing situations is competence. We organize that little bit of the situation which we are able to see and fit it to the projects we are capable of formulating. For example, for those who understand French but not English, only a small part of this work will be accessible, thanks partly to the words with Latin roots that are common to both languages, but mainly to the fact that both languages use the same set of characters.

For those who cannot read any language but who know that writing and books exist, this book is an object containing sheets of paper covered with indecipherable marks. For those who not only cannot read but who also do not know what writing is and have never seen a book, this volume would be an alien and incomprehensible object. It would seem strange, these days, to imagine a person who has never seen a book. However, the above example emphasizes the great number of competences, often implicit and therefore taken for granted, which we use almost without realizing exactly how many there are, to structure situations in everyday life.

History and cultural anthropology supply us with abundant illustrations of similar situations. More than four centuries ago, in 1532, at Cajamarca in Peru, the Spanish *conquistadores* captured the Inca king Atahualpa and put a copy of the Gospel into his hands. But he let it fall to the ground, horrified and disoriented. Presumably, he did not imagine what those sheets covered with little black marks were: for him, the foreigners were violating the solemn taboo that prohibited contact with his person. He might well have supposed that the marks represented some kind of sign but, in order to be sure, he would have had to know of the existence of writing and, if possible, known those particular characters and the language which would have allowed him to read the text. However, even if he had been able to read, he would have found the Gospel completely incomprehensible. Not only would he not have understood the text, telling strange stories of deities incompatible with those he knew and duly represented, but in particular he would not have been able to grasp what the Spanish required of him when they thrust this weird object into his hands. In order to extract some plausible meaning from the situation, he should have resorted not to the book itself but to the Spaniards, who would have explained why they had pressed it upon him. They would inevitably have had to refer to their own intentions, stating, for example, that they had given him the Gospel to justify his capture with the accusation of impiety, or that they wished to protect him, if he had been 'converted'.

In any case, after his initial surprise and dismay, Atahualpa doubtless constructed his own explanation of the situation. In this he was helped not so much by words but by actions: the Spaniards instantly seized him and demanded an immense ransom in gold. It was probably at that moment that the Inca king was obliged to realize his inability to handle the unforeseen situation. His ignorance of the Gospel was only an indication of the greatest and, for him, disastrous incompetence in his encounter with the Europeans. He knew too little of the cultural scenario in which his captors moved and, fatally, was unable to cope with it.

In order to see whether our construction of a situation requires competences, either specific or general, we do not necessarily have to resort to episodes as remote in space and time as the above example. Even in order to find mushrooms in a wood we have to know what looking for. We have to know what perceptive indications can lead us to infer the probable presence of mushrooms and the appropriate actions for collecting them. We do not just perceive or know things in a *vacuum*: 'we see, hear and feel things in a context, against a background of other things and actions. Our perception is embedded in sources of action which are themselves patterned, learned and shared. Objects, processes and actions are set against one another in that they emerge from a retreat to a background. The environment is structured so that feeling, hearing and seeing is possible' (Anderson and Sharrock, 1993, p.146).

Although the selection of information to be extracted from the environment for specific purposes can require specific cognitive competences, this it is not necessarily a conscious and attentional activity. There is solid experimental evidence of learning and problem-solving processes which remain unconscious and automatic even though they are complex and highly sophisticated, e.g. decision-making in the sphere of finance and business (Hayes and Broadbent, 1988).

There is no information available in general, but actors seek, interpret and use all kinds of clues to develop their knowledge of the environment and their actions in it: 'information is created by the observer, because comprehending is conceiving, not retrieving and matching' (Clancey, 1993, p.91). Perception and knowledge of situations are linked to action: 'Learning is inherently "situated" because every new activation is part of an ongoing perception–action coordination' (*ibid.*, p.95).

1.3 AMBIGUITY OF SITUATIONS AND INCONSISTENCY OF INTERESTS

The second factor that enters the construction of situations is motive. In order to organize a situation in a certain way, we must see that it contains some aspects which are attractive or repellent to our interests. Identification of the promises or threats which a situation contains for a particular actor in a particular space–time unit depends not only on competence in understanding the potentially relevant aspects of the situation, but also on the presence in that actor of interests which are sufficiently developed and articulated to be recognizable, at least by her/him, and are therefore made salient in that situation.

In opening their windows on situations, actors are motivated by their goals. A situation may be conceived as lacking in hints not because of its inherent poverty, but because of the actor's poverty of interests. A situation which may appear lacking in significant development to a less competent or less motivated actor, may appear loaded with promise to one who is more expert or imaginative. A promise (or a threat) is one aspect of the situation which appears to be favourable (or adverse) to the achievement of one of the actor's explicit or latent projects. The fewer interests and projects an actor has, the fewer opportunities he will find in the environment: those who have many interests will find even more of them; those who have fewer will see that even what they have will be taken from them.

The construction and interpretation of a situation imply assessment of the appropriateness of the transaction between actors and their current environment. This is not assessment of the situation in itself, but of how it fits, or does not fit, into the idea actors have of their goals. In the same space–time unit – let us say, a lesson – in which an actor X sees an interesting opportunity to learn something new in physics, another actor Y sitting next to him sees an equally exciting opportunity of making the acquaintance of actor Z, who has long blond hair and an enchanting smile. These attributes are salient aspects of the current situation only within the framework of Y's interests at that moment.

The above situation is not perceived identically by X and Y, even though they are sitting next to each other in the same classroom at the same moment. In a strict sense, they are experiencing two different situations organized by two different systems of interests. It is not only the situation but also the attitude of the actor which requires adequate diagnosis: the actor who finds himself in what would be a promising situation for an up-and-coming new writer may miss the chance offered him, because he does not know that he is, or may become, an excellent writer.

In our example, X follows the lesson while Y flirts with Z. Each actor has only one interest and only one behaviour. However, things are not usually so simple, because in reality X and Y almost always do several things at once, some of them being easily visible while others are less apparent – as is the case, for example, of daydreaming or thinking. We therefore see that, while X listens to the teacher and takes notes, he can also adjust his glasses, pull at his hair, and loosen a shoe. Still following the lesson, X also remembers that on Sunday he plans to go skiing, that he must not forget to call Anne and make humble apologies, ask Y to lend him some money, and so on. This is because X, like Y, has various interests at the same time.

Interests sometimes appear to be compatible, with regard to both the types and the extent of the requests made on X's attention system. In these cases, more or less easily and with some degree of compromise, X manages to satisfy all of them. In other, quite common cases, X's interests are either in conflict or incompatible with the demands they make on his system of attention and action. The idea of snoozing comfortably in bed until late on Sunday is incompatible with leaping out of that nice warm bed at the crack of dawn to go skiing. In X's cultural context, which forbids polygamy, an interest in marrying Anne is incompatible with an equally strong interest in engaging in a love affair with Jean.

We have listed some of X's many possible interests in order to justify the continuous oscillations in the direction and intensity of an actor's attentional focus. We all know that everyday situations are in fact far more complicated than our examples show. To approach daily situations, we must think of the plots of such classic novels as those of Balzac, Tolstoy or Dostoyevsky; the stories may span years, like *War and Peace*, or even generations, as in *The Brothers Karamazov*, and are brimming over with despair, sorrow, joy, greed, compassion. We can find everything in daily life except perfect consistency.

Let us now imagine X's attentional system as if it were the cash window of a bank where a long line of creditors – all more or less vociferous, argumentative and justified in their requests – are hammering on the counter asking for their money back. If we presume that the system of competing interests, the creditors, tends to claim more cash than the bank can make available, dispute over access to resources such as attention and action may easily degenerate into open conflict.

In the arena of the actor's attentional and decisional space, those with the best justifications or rights who can shout loudest and longest tend to be satisfied first (Payne *et al.*, 1992). It is not out of place to speak of rights in this case because, as we shall see later *à propos* image theory in intuitive or automatic decision-making, interests are highly sensitive to norm systems and quite often stake claims precisely in terms of values and principles (Beach, 1990).

Actors, unless they are suffering from some severe psychic disorder, do not have just one interest at a time; nor do they develop a perfectly stable and clearly ordered scale of interests. If a situation at a given moment is interpreted in a certain way, this is not necessarily because we do not see other ways of interpreting it, but because the way in which we choose to interpret it prevails over the alternatives at that moment in time.

This is why our responses to everyday situations are often so unpredictable. It is difficult to account not only for differences among actors, each with their own goals, but also for the changes in priorities among a single actor's interests. At a party, for example, not only do X and Y differ in their actions and in their interpretions of the event, but X may change his goals, mood and evaluation of the situation several times during the evening. He may do this without being aware of his motivational state, at least at first, or without being able to control it. He may become gloomier and gloomier or more and more cheerful, for reasons which partly elude him and which may have nothing to do with the execution of any deliberate plan.

So it is not an informational overload which lies at the root of the complexity of everyday situations, since in any case humans select and process the information they need at various levels. The true complexity of daily situations – which is responsible for the fact that they do not yield to treatment by analytical tools and resist reduction to formalized predetermined models – comes from the multiplicity of actors' conflicting interests and from the brilliant cognitive resources which they bring to bear at any given moment in order to achieve their current goals.

This fact becomes particularly evident in encounters between people from different cultures, although the primary actors may not realize it clearly: the situation faced by Atahualpa was not the same as that faced by Pizarro. The Inca king saw sacrilegious foreigners touching him with a strange object and treacherously capturing him while he was paying a courtesy visit to their camp. Pizarro probably saw his group of brave warriors as being ready to seize a golden opportunity to capture the person of an absolute monarch, to wipe out the enemy forces, and to ensure the overwhelming success of their enterprise, to the greater glory of God.

In cases like this, of cultural (and often not only cultural) conflict, there are clearly two different situations, without possible mediations. Both Atahualpa and Pizarro constructed and interpreted events according to the cognitive and motivational resources available to them. There was no neutral backdrop, no 'true' reconstruction – in the impartial and exhaustive sense – of events. Today, German and French history books tell the history of the First and Second World Wars in quite different ways. Neither side is lying, both countries have excellent historians, and the good faith and professionalism of academic scholars and school teachers are beyond question, and yet – or perhaps for this very reason – they tell two rather different stories, giving contrasting interpretations of events.

1.4 PARTICIPATORY INTERACTIONS BETWEEN AGENTS AND ENVIRONMENTS

It is the instability of the actor's system of interests that continually questions which criteria are important for the situated actor's information selection. If I am going to work, am late, feel in need of something to drink, and see my friend David, I can either slap him on the shoulder and go off to drink something with him at the nearest café – thus ordering priorities among my interests of the moment, since punctuality now becomes of secondary importance – or I can simply wave to him, hurry on, and get to the office on time, thus establishing another order of interests in which punctuality overrides the desire for a drink and a chat with David.

Whatever my choice, the fact that I have to act in one way or the other, hurrying on or stopping to chat, means that I am obliged to order my system of priorities and make it at least temporarily consistent, so that the situation is granted a precise, albeit precarious, framework. The action provisionally structures the situation, shielding it from ambiguity for a moment. Ambiguity again rears its head when, a moment later, I reconsider the consequences of my action in another context and perhaps give them another interpretation. If I stop at the cafe with David and get to the office late, I may regret my decision later, when the consequences of my action turn out to be less pleasant than anticipated.

An action that appears suitable at one moment may later turn out to be inappropriate, not because the means–end ratio has changed, but because the actor's goal has changed: earlier I wanted a drink and a chat with my friend, now I regret that I am late for work. It is this continually changing transaction between occasions offered by situations and the *hic et nunc* salient interests of the actor which makes interpretation of the situation precarious. It is the changing nature of our systems of interests which is responsible for ambiguity, because interests construct situations, identifying in them promises or threats which are only such in view of their relevance to the implementation of our goals (Frijda, 1986, 1987; Frijda and Swagerman, 1987; Lazarus, 1991). Situations in turn interfere with systems of interests, favouring some at the expense of others.

Research on communication and cooperation stresses that situations are also complicated by the number of actors participating in the interaction (Schmidt, 1991), each with their own idiosyncratic vision of the world which is the result of particular competences, practices and lexicons (Williams and Gibson, 1990). However, we know that situations are complex even when only a single actor is on the stage, because even the goals of one person may be disordered and inconsistent.

Speaking of disorder in human cognitive and motivational functioning may be misleading, if we think that actors move at random, unable to cope with the requirements of their surroundings. This is not the case; what may seem to be a defect is in fact an extraordinary capacity for adaptation to the environment and its changes. The human system of interests, thanks precisely to its instability, allows flexible and effective responses to a vast range of environmental scenarios. Disorder and even inconsistency among interests prompts actors to change their plans and to redefine their priorities, both in relation to the results of their actions and following sudden transformations in the environment or in the system of principles to which they refer (Beach, 1990; Mitchell and Beach, 1990).

It is difficult to acknowledge the ambiguity of both situations and actors inside a scientific discourse on social or cognitive psychology. However, it is precisely the relative maturity of these fields which makes this approach feasible today: 'The resources that are available now in ecological psychology, ethnography, and philosophical situation theory seem to us to provide a prospect of progressing substantially toward a rigorous and detailed analysis of cognitive processes considered as participatory interactions between agents and physical and social systems' (Greeno and Moore, 1993, p.58). Greeno and Moore see a new frontier in psychological research in the acceptance of everyday situations. Yesterday, they say, behaviourism forbade the 'black box' containing mental processes to be opened. It was later opened thanks to the theory of symbolic processing. Today we are faced with another 'black box' containing the complex structures of daily interactions between agents and the environment. We can now try to open it with the aid of the psychological, ecological and ethnographic theories, all of which flow into 'situativity theory'. We may also realize that the new black box need not be forced open because it was never really closed. To Vera and Simon (1993, p.45), who see real-world situations as complex from the cognitive viewpoint, Suchman (1993) replies that 'the complexity or simplicity of situations is a distinction that inheres not in situations but in our characterizations of them. All situations are complex under some views, simple under others' (pp.74–75). This ethnographic observation fits perfectly with recent research results in the field of social psychology.

The general aim of this discussion on actors and situations focuses attention on the area in which, in Figure 1, the *actor* and *environment* systems overlap. It defines the space of the actor–situation relation: on the one hand, the actor's selection processes; on the other, the promises or threats latent in the situation. In the area of overlap, which is where everyday experience in fact takes place, we cannot clearly separate the actor from the situation. The interaction is so close that the actor-in-situation is defined precisely by the way in which he exploits opportunities. Conversely, the situation does not exist before the actor enters, but is constructed by his intervention in it: this is why X's physics lesson was different from Y's flirting session.

If we refer to what happens in daily experience, the actor–environment situation cannot really be adequately explained by metaphors (to which we are only too readily inclined) based on the separation between internal mental processes and external events. Much of our current knowledge resides in the world, not inside our heads, as Norman (1988) and Zhang and Norman (1994) maintain when speaking of artifacts. In order to understand everyday experience in its entirety, we have to overcome the dichotomy, inherent in common sense and most current behavioural research, between purely physical things outside our heads, like apples, hammers and cars, and things inside them, like knowledge, ideas and plans.

We concur with the insightful statement of Agre (1993): 'My point is that the metaphor system of Inside and Outside finds itself unable to make any stable sense of concepts that reside neither in the agent nor in the world, but in the relationship and interaction between the two' (p.67). Everyday experience mingles Inside and Outside: at the beginning of *À la Recherche du temps perdu*, are the smell of the *madeleine* and the memories it awakens outside or inside Proust's mind? Memory, perception and emotions appear to be more accessible through metaphors of the key–lock type than on the basis of conceptions which consider human beings as if they were silhouetted against an indifferent, extraneous or even hostile environment.

CHAPTER TWO

Circumstances, Plans and Situated Action

The term 'situated action' underscores the view that every course of action depends in essential ways upon its material and social circumstances. Rather than attempting to abstract action away from its circumstances and represent it as a rational plan, the approach is to study how people use their circumstances to achieve intelligent action. (Suchman, 1987)

What traditional behavioral sciences take to be cognitive phenomena have an essential relationship to a publicly available, collaboratively organized world of artifacts and actions. (*ibid.*)

2.1 OPPORTUNISTIC STRATEGIES AND CONSTANCY OF GOALS

We have seen that actors' systems of interests are often disordered and that environments tend to be unstable. Is it therefore impossible for actors to develop long-term projects and persevere in their attempts to realize them? Recognizing the widespread presence and adaptive value of 'opportunistic strategies' may lead us to believe that people, groups and organizations simply wander about in daily reality without stable goals, prey to fluctuating moods and to the continually changing situations which surround them.

However, actors are not completely dominated by opportunities that could be grasped. Nor do they seem to be lacking in firm ideas about the directions in which they should move, or to be incapable of persevering with their goals. Competent social actors are not at the mercy of their social and physical environment like ships battered by storm waves, content just to keep afloat for the time being without worrying about reaching port. Agre and Chapman (1987, p.268) state that: 'Before and beneath any activity of plan following, life is a continual improvisation, a matter of deciding what to do now based on how the world is now'. We cannot subscribe wholly to this view, since reality presents us daily with cases of actors who persevere mightily in pursuing their goals even in unfavourable circumstances and in permanently adverse conditions.

Improvisation in human conduct usually exists at lower but not at higher levels of planning. It exists at the level of short-term action plans, which are adapted almost automatically to particular circumstances. For example, if I want to invite Sally to a party but see that she is in a bad mood, I can wait for a better moment; in the case of a bus strike, I can go to school by bike; and if my bike has a flat tyre I can ask Andrew for a lift. On the level of small segments of daily planning, the rule is maximum flexibility, and actors are normally quick at finding on-the-spot solutions to problems as they arise.

But improvisation can hardly exist at the level of medium- and long-term plans. If I am really determined to be a psychologist when I grow up, the fact that the university in my home town does not have a faculty of psychology will not deter me or induce me to enrol in some other faculty. In order to reach my goal, I will simply decide to study somewhere else, even if this means sometimes considerable hardship. If Maureen wants to live with David, she will not leave him for someone else simply because he has to work abroad for a couple of years; on the contrary, she will probably make plans to join him, excogitating subplans about how to get hold of the money for the air fare. People are capable of great determination when fighting to maintain principles or reach goals that are really important to them; in these cases, they can stay true to prior purposes even for long periods of time and in highly unfavourable circumstances.

Distinguishing between more or less important goals allows us to reconcile the instability of situations and of actors regarding the most circumscribed questions with their well-known persistence on important issues. Clearly, opportunistic strategies do not hinder the success of medium- and long-term plans; they favour them. And it is precisely this flexibility in short-term planning which allows medium- and long-term plans to be developed successfully.

In everyday life, something similar to the Italian *commedia dell'arte* of the sixteenth and seventeenth centuries occurs. The actors (understood here as stage actors, of course) were highly talented professionals who inherited their skills from the Medieval jesters, clowns and acrobats, and they had to be able to follow scripts that only roughly indicated the development of the plot. The scripts of the *commedia dell'arte* which have come down to us contain only vague, brief stage directions such as 'Tritone enters, squirts water over everyone, and leaves', or 'Arlecchino enters, cuts a few capers, and leaves', on the basis of which the actors had to improvise scenes sometimes lasting up to ten minutes. This was possible because they filled in the gaps with songs, miming, acrobatic displays and music, at all of which they had to be adept. The ability to adapt ourselves to circumstances is still essential in the difficult, tiring, socially demanding but also gratifying duty to perform as adequate and competent actors on the social stage.

However, the great flexibility of conduct required of the *commedia dell'arte* actors during their performances on stage (which we interpret as short-term action plans) corresponded to extreme rigidity of characterization (medium- and long-term plans), which was taken on by various personages. While Arlecchino's buffoonery, Dr Balanzone's rambling speech and Captain Fracassa's bravado could all be infinitely varied, the characters themselves were not allowed to change. They were fixed, ideal types of figures of the society of those times: the shrewd but idle servant, the learned, pompous lawyer, or the arrogant, boastful soldier. Daily experience, like the *commedia dell'arte* scripts, requires actors to link improvisation in the management of single scenes with the permanent figure of the character, i.e. personal identity and the aims most closely connected with it.

The popular metaphor of life as a theatrical stage guides us here: improvisation inspires Arlecchino's movements and cues on the stage while his mask and character are fixed. In each company of actors, it was always the same person who played the part of Arlecchino. Character stability and improvisation on the stage were harmoniously coupled in the *commedia dell'arte*. The more the former was relied upon, the more the latter appeared as simple and natural. The more skill and specialization lay behind the mask, the more the actor's behaviour on stage was relaxed and apparently casual.

Improvisation occurs when ways and means can be easily exchanged without changing goals, or when goals are of limited importance for the actor, so that the means–end relationship may be quickly reformulated or even reversed to fit the situation. For example, if Brian wants to go to class, this means either taking the bus or using his bicycle, because he lives out in the suburbs. It so happens that neither form of transport is available today: no buses, because of a public transport strike, and no bike, because it has no lights and he will be back after dark – and, come to think of it, its brakes are pretty bad too. ('Why not use the car, just this once?' 'Oh, yeah, very funny, do you *know* how much it costs to have wheel-clamps removed?'). So Brian gives up his original idea of going to class today, unless he is really obliged to do so.

Conversely, improvisation does not occur when actors' more general goals are involved – goals involving their principles and identities. If I really want to be a psychologist, I will not be discouraged by sometimes serious difficulties that may arise along the way. If Maureen really wants to live with David, she will not readily give up opportunities of being with him; on the contrary, difficulties may even increase her determination: difficult or impossible loves are, not by chance, really irresistible.

In pursuing goals that are hindered by somehow important situations, people not only reveal the determination of their commitment towards those goals, but also state their relative autonomy with respect to contingent situations. As in nineteenth-century love stories and romances, they overcome contingent difficulties by faithfully following a scenario based, on the one hand, on personal choices and identities and, on the other, on a more general context which may reshape immediate contingencies and may sometimes subvert the initially unfavourable situation. The stock scenario is that, after many adventures and tests of endurance, the heroine's virtues are finally recognized and the villains are unmasked and punished. Stories like *The Count of Monte Christo* tell us that the hero can unexpectedly escape from a heavily guarded prison and take revenge on his enemies, if he shows himself capable of coping with present difficulties without giving up hope.

However, it is sometimes difficult to make a clear-cut distinction between 'local' goals, which are more sensitive to situations, and general goals, which are less easy to negotiate. One instrument that allows this distinction is image theory (see Chapter 3), which links the maintenance and development of personal and social identity with a special decision-making model. In any case, demarcation between local, only instrumental goals, and general goals, which are important for actors' identities, is difficult because the relationship between goals and an actor's identity defines a continuum rather than a dichotomy. In effect, there are no goals which, by their very nature, are unimportant for actors; we can easily imagine a situation in which it is more important for Sharon to remove an oil stain from her smart new suede jacket than it is for her to pass a statistics exam.

2.2   ENVIRONMENTS, GOALS AND THE DEVELOPMENT OF IDENTITY

In daily life, our goals come into being and are developed both in a conscious, deliberate way, as a consequence of direct investment of attention, and in an automatic, unconscious way. Lazarus (1991) speaks here of preconscious or even unconscious processes, which we accept in as far as we must not presume that they are in principle inaccessible to conscience, being, for example, repressed. Since allocation of attention to special aspects of the environment is unavoidably unstable, intermittent and selective, we may suppose that actors' goals occur in a mainly automatic way in daily reality.

To the extent to which the definition of goals is entrusted to automatic processes, situations – like our memory of them – have a considerable influence on the generation of goals. Situations, and our memory of them, take on configurations that are capable of activating special sequences of motives–goals–plans in actors. When Sharon returns to class after having been away for a few days with flu, her friends, without needing to reflect in any particular way, exhibit a series of behaviours which are appropriate to the situation. They gather round, tell her the latest news, lend her their notes, etc. 'The result of this automatic associative link is for the motive–goal–plan structure to become activated whenever the relevant triggering situational features are present in the environment. The activated goals and plans then presumably guide the social cognition and interaction of the individual, without the person's intention or awareness of the motive's guiding role' (Bargh, 1990, p.100).

Situations do not only offer elements suitable for generating sequences of goals and activating special action plans, which are not necessarily consciously controlled by actors, but also contribute significantly to the formation and development of self-identity: 'The self-concept is not present at birth but arises out of social experience and interaction; it both incorporates and is influenced by the individual's location in the social structure; it is formed within institutional systems, such as the family, school, economy, church' (Rosenberg, 1981, p.593). The self-image is made up of the materials of a given culture and is sensitive to the social and environmental contexts that surround it.

The process of self-construction has its roots in the circumstances of daily social exchange, as Mead noted as long ago as 1934: 'The individual enters his own experience as a self or individual not directly or immediately, but only insofar as he first becomes an object to himself just as other individuals are objects to him or in his experience; and he becomes an object to himself only by taking the attitudes of other individuals toward himself within a social environment or context of experience and behaviour in which both he and they are involved' (p.138).

In socialization theories, personality is inextricably linked to the social circumstances, to the extent that, by the concept of personality, we mean that particular configuration of attitudes, motives and competences which make up the specific response of a given individual to the requirements of the environment (Hurrelmann, 1988). Research, especially in the field of social psychology, confirms the dependence of self on the social and physical context (Markus and Kunda, 1986; Markus and Wulf, 1987; Gergen and Gergen, 1988; McGuire and McGuire, 1988).

If we place ourselves in a situational perspective, the link between environment and self-development appears to be quite close. Going back to the metaphor of the key and the lock, Baron and Boudreau (1987) state: 'Personality and social psychology are inextricably bound, with the major difference being that personality theorists are oriented more toward specifying the nature of keys, whereas social psychologists focus on specifying the nature of locks' (p.1227). From this viewpoint, the most precious aspect of personality, its uniqueness, appears as the reverse of the coin of situational specificity. Every personality is unique because every actor fits his or her own lock in a way that no other actor can precisely reproduce, and acts as a lock in an inimitable way for a certain clearly defined range of other actors.

Personality and environment are complementary. The former is a set of predispositions allowing actors to identify certain affordances in the environment and to respond suitably. The latter sends out a series of signals which select the actors interested in them because they are able to understand or are motivated by that particular affordance. We cannot think of goals as being 'inside' people and situations as 'outside' them. The persistence of goals is the result of an encounter between certain actors' permanent needs and certain corresponding characteristics available in the social and physical environment.

2.3 CIRCUMSTANCES AND PLANS IN PURPOSEFUL ACTIONS

Although we may be inclined to acknowledge that opportunistic strategies do not stop actors from developing their own long-term plans in view of relatively permanent and autonomously defined goals, common sense constantly puts us at the risk of falling into that dichotomy between subject and environment, actor and scenario, 'inside' and 'outside' people's heads, which we criticized at the end of Chapter 1. Common sense suggests that the origin of goal stability can be traced to an actor's will and determination, in contrast with the instability of situations.

We are used to believing that goals reside inside people's heads, since it is difficult to avoid a way of thinking and speaking which constantly counterposes subject and situation. Clearly – we are led to assume – if actors are capable of producing lasting plans, this is because they are capable of resisting the pressures of a confusing and distracting environment, thus preserving their goals. Just as Ulysses was only able to resist the fascinating song of the sirens by having himself tied to the mast of his ship, so actors can remain faithful to themselves only by closing their eyes and ears to the flattery inherent in situations. This is what we believe, but in fact we should ask ourselves: does Ulysses' love for his island home, for his wife, for his father, only exist inside him, purely as a mental event, or does it also dwell in Ithaca, Penelope and Laertes? Do these bonds come into being and develop only 'inside' Ulysses, with no 'external' contribution by situations and other people, or are they things which his island, his wife and his father contribute toward creating and nourishing, and do they therefore belong to them as well as to him?

If we opt for the latter alternative, we view Ulysses' love for Ithaca as an encounter that occurs in that area in which the events of the environment and of the actor overlap and become one thing. The desire to return to Ithaca belongs to Ulysses, but Ithaca is not extraneous to this goal; in a certain sense, it is a common goal. Lazarus (1991) highlights this point, which is as essential here as it is in his relational and appraisal theory of emotions: 'If we feel threatened, insulted or benefited – these are, of course, appraisals – the relational meaning of each does not stem from either the person or the environment; there must be a conjunction of an environment with certain attributes and a person with certain attributes, which together produce the relational meaning' (p.90).

We find it difficult to reshape our cultural and linguistic categories taking into account the relational nature of the actor–environment interplay. At the moment we really think about the idea of relation, it becomes necessary for us to resort to a language capable of considering the two complex subsystems – person and environment – jointly, as they come together in their interaction. The separate identities of the two subsystems become blurred, giving way to a new condition that has to be described in new relational terms.

It is in this area of ongoing actor–environment interaction that goals are formed, not in the isolation of an individual mind, understood as something opposite to the social and physical environment. The relational meaning of a situation is not a moment of confusion or conceptual approximation. On the contrary, it institutes a level of study higher than that consisting in analysis of the individual dimensions which produced it. The social and behavioural sciences still have difficulty in accepting this approach. It not only involves a higher level of abstraction; it also requires effort aimed at joining up all the complex elements composing the relation itself. 'This seems to be a difficult idea for modern social scientists to grasp, perhaps because they have been reared professionally to venerate analysis and the partitioning of variance – as in analysis of variance – as the ideal model for scientific understanding' (Lazarus, 1991).

Inevitably, opposition between actor and situation continually arises, in both common and scientific discourses. It reflects a characteristic of Western culture: that of thinking of order and rationality as marks to be branded by the mind on 'external' environments which would otherwise be chaotic and insensate. This viewpoint has come to be part of common sense and also appears in the semantic components of discourse. If we speak of goals, we automatically think they can only exist inside people's heads. We are led to believe that the reason for the stability of Maureen's goal to live with David is to be sought inside Maureen; that it is one of her personal characteristics. It is in fact the result of the transaction between certain characteristics of the situation and certain characteristics of Maureen herself.

To explain the origin of difficulties found when trying to reconcile adaptation to situations and permanence of goals, we have to consider which conceptual categories we use when thinking of goals and situations. We must realize that they depend on a given culture, that they are not absolute and applicable to all human beings, groups or organizations on the face of the earth (Archer, 1988; Sahlins, 1985). Until now, psychology has continued to decontextualize individuals, isolating them from their social and cultural environments (Cushman, 1990), but things begin to change if we peer inside the new 'black box', the one containing the actor–situation relation.

In order to capture the meaning we give to goals, plans and actions, it may be useful to quote a passage from Berreman (1966) – which in turn comments on Gladwin's paper (1964) on the Puluwat navigators of Micronesia – placed at the very beginning of Suchman's book on plans and situated action (1987). Berreman compares the Western theories and practice of navigation with those used by the natives of the island of Puluwat, the Trukese:

> The European navigator begins with a plan which he has charted according to certain universal principles, and he carries out his voyage by relating his every move to that plan. His effort throughout his voyage is directed to remaining 'on course'. If unexpected events occur, he must first alter his plan, then respond accordingly. The Trukese navigator begins with an objective rather than a plan. He sets off toward the objective and responds to conditions as they arise in an *ad hoc* fashion. He utilizes information provided by the wind, the waves, the tide and current, the fauna, the stars, the clouds, the sound of the water on the side of the boat, and he steers accordingly. If asked, he can point to his objective at any moment, but he cannot describe his course. (Berreman, 1966, p.347)

The difference between proceeding towards objectives with continual adjustments, like the Trukese, and following a pre-existing analytical plan, as Westerners do, sums up some of the observations made up to now and also introduces two new questions: what relationship is there between plans and action? Can a coordinated sequence of actions be developed without a previously established plan? We will answer these questions in the next section.

Before examining them, we must ask ourselves whether we are sure of the fact that the above-mentioned Trukese and Western habits are two ways of navigating or whether they are two ways of *thinking* about navigation. In the first case, we have two different lines of action, each effective in its own way. In the second, we have two different ideas about navigation, which may not necessarily correspond to two different systems and practices of navigation on the high seas. When we ask ourselves if the two reports effectively correspond to different strategies, what we are really doing is entering the complexity peculiar to daily situations, in which we cannot clearly distinguish data from interpretations.

We can say that Westerners and Truk natives have developed two ways of thinking about navigation which aim at controlling the corresponding practices in a more or less comprehensive manner. European culture has preferentially developed forms of abstract thought, consisting in starting from general principles and only later going on to definite, particular applications. Instead, Trukese culture directly supplies responses flexibly adapted to the situation and incorporated in the actor, his instruments, and the physical and social environment. If we liken nautical skills to a cultural construction, we might expect Trukese and European cultures to develop two different ways of navigating.

But things do not always develop like this. According to Suchman, Westerners think they act in a certain way but in fact they do something slightly different. They think they are following predetermined plans but in reality they too are constantly adapting their plans to situations: 'However planned, purposeful actions are inevitably situated actions. One could argue that we all act like Trukese, however much some of us may talk like Europeans' (*ibid.*, pp.viii–ix).

In his review of Suchman's book, Agre (1990) comments that such complex activities like those involved in ocean navigation always require careful preparation and appropriate planning. He quotes a passage from a further study by Gladwin (1970) that rectifies the incorrect idea that Trukese navigators act in a planless way:

> The Puluwat navigator also has advance plans which cover the entire voyage. These are the sailing directions he has learned and they are quite as complete as those of any Western skipper. The difference is rather that the Puluwat navigator has his plans available before the voyage is even prepared. He has had them ever since he · learned navigation. The Western navigator in contrast makes up a new one for each trip. Thus we come again to the matter of innovation. Yet by the time both the Western and the Puluwat navigators are ready to get under way their plans are remarkably similar. They are based on somewhat different maps, both cognitive and on paper, and the process as a whole seems superficially very different, but they cover the same things for the same reasons. (Gladwin, 1970, p.232)

The reason why purposeful actions must be negotiated with the environment through flexible planning is that situations cannot be entirely reduced to actors' capacities for prediction. Contingencies as they present themselves in daily experience are always unpredictable from certain points of view, so that actors, if they wish to continue to steer towards their goals, have no alternative but to adapt their plans to circumstances as smoothly as possible. Only a helmsman not constrained to steer towards a precise goal can apply his plan inflexibly: once he has chosen his direction and fixed his rudder, circumstances will inevitably take him *some-where*. But a navigator who is obliged to reach a certain port at a certain time has to go to great lengths to anticipate and compensate for all contingencies which might unexpectedly take him off course or delay him.

The partial unpredictability inherent in every situation appears to be the grain of sand in the machinery of pre-established plans considered as calculation procedures. Suchman's advice is: 'We must act like the Trukese because the circumstances of our actions are never fully anticipated and are continuously changing around us. As a consequence our actions, while systematic, are never planned in the strong sense that cognitive science would have it' (1987, p.ix).

2.4 PLANS: PRESCRIPTIONS OR RESOURCES FOR ACTION?

Cognitive science, according to the perspective of situated action, overturns the action–plan relation, assigning to plans the task of controlling and checking actions to be undertaken, instead of functioning as resources for action, repertories of possible responses to be activated in order to cope with specific contexts and problems. In this reversal of perspectives, what is lost is the adaptive character of the relation between actor and context. Conversely, if we view planning as a situated process (Rogoff et al., 1994), action becomes naturally adapted to the environment precisely because it starts from circumstances and organizes itself so as to exploit them to the full.

The perspective of situated action does not consider irrelevant the particular social and physical circumstances in which social actors are called upon to operate, and does not attempt to ignore them in order to impose a pre-established order on events. Every intelligent action which is adapted to its environment resorts to plans, but they are plans corresponding to the course of the action and interpreted with reference to it (Mantovani, 1996). Plans lie within situations and within the actions that interpret them, and not the other way round, which has been the approach followed by cognitive science starting from the work of Miller et al. (1960). Situated action does not come into conflict with the symbolic approach traditional to cognitive science; rather, it enriches it (Clancey et al., 1994) by suggesting that plan construction could be viewed as a real context-dependent activity. Vera and Simon (1993) on the one hand, and Agre (1993), Clancey (1993), Greeno and Moore (1993) on the other, although with different accentuations, agree on this point. We realize that the current concept of plans as rational organizers of action is mainly a cultural artifact that gives a substantially misleading vision of the interaction between actors and the environment: 'It is only when we are pressed to account for the rationality of our actions, given the biases of European culture, that we invoke the guidance of a plan' (Suchman, 1987, p.ix).

It is not correct to say that plans always precede action and entirely control its course. They are often reconstructions, established and negotiated after the action, by means of which social actors try to explain what has happened according to their available cultural categories. Criticism of the reconstructive nature of plans is implicit in the discussion on the validity of verbal protocols. This is contested by Nisbett and Wilson (1977), who believe that people cannot completely and accurately grasp the relation existing between stimulus and response, even when both are known. To an even greater extent, therefore, verbal reports are of little assistance in the cases of unconscious and automatic learning studied by Hayes and Broadbent (1988). Ericcson and Simon (1984) reply by partly reassessing the validity of verbal protocols, at least in cases in which the report is close in time to the reconstructed event, is not guided by the researcher according to predefined categories, uses visual codes, and operates in the ambit of short-term memory. The research technique consisting of asking subjects to 'think aloud' while they perform a task is presumed to supply immediate and therefore reliable reports, seeing that, in simultaneous reporting, distorting factors attributable to mediated processes are not assumed to be introduced between the event and its verbalization.

Considering the inevitably interpretative character of situations, the solution proposed by Ericcson and Simon does not appear to be very satisfactory. Even if we admit that there is no time lag between the event and its verbal reconstruction (in fact, of course, there *is* a delay), any representation which subjects construct of environments and the course of events must be influenced by socially mediated filtering processes. In reality, actors do not limit themselves to being present in situations, but contribute actively to their own constructions and interpretations. We can see here the implications of our relational approach: while Ericcson and Simon believe that situations are known only partially, we believe that they are accessible by means of interpretative work.

Cognitive research has emphasized the incompleteness of available information and lays the blame on the limitations of short-term memory, considered as the main drawback to the development of full rationality in human decision-making (Simon, 1981, 1983). Instead, we prefer to see situations not as known in an incomplete or defective way, but as constructed and interpreted – within the framework of a given culture – in a way that is more or less appropriate to actors' interests and capabilities on the one hand, and to the opportunities afforded by the situations themselves on the other. It is within this framework that we can appreciate the full significance of the perspective of Norman (1988; Zhang and Norman, 1994), which recognizes the intelligence quotient lying inside actors' heads, inside things, and inside the way in which artifacts structure the information they make available to actors. A display may be decisive in allowing a critical piece of information to be acquired or in processing a complex situation properly. Competences may even have an extra-somatic existence, as is increasingly occurring in the era of 'cognitive artifacts'.

As we conceive of situations as transitions between actors and the environment, we have no difficulty in accepting the presence – in our concepts of plans, goals and action – of the categories proper to the social and cultural context in which we live. But traditional cognitive science tends to neglect the intervention of these categories and to present its own conceptualizations as absolute data, valid independently of particular social and cultural contexts. By means of the idea that plans control action and, through it, the social and physical environment, cognitive science credits the image of a rational order of the world instituted by the mind and imposed by the mind on a reluctant and chaotic environment. In such a vision, the real interaction and overlap between actor and environment is hidden.

Situated action questions the cultural significance of the reversal of priorities which cognitive science operates between actions and plans: 'My aim in offering an alternative position was not to argue the nonexistence of plans (such a position would be *prima facie* absurd) nor their irrelevance to human action. My aim was to raise a basic question about the *status* of plans, both as artifacts of the situated activity of projecting future actions, and in relation to the actions they project' (Suchman, 1993, p.73). In the area of research and application known as *computer-supported cooperative work* (CSCW; see Chapter 6), we consider the increasing acceptance in CSCW studies of the principle according to which 'what traditional behavioural sciences take to be cognitive phenomena have an essential relationship to a publicly available, collaboratively organized world of artifacts and actions' (Suchman, 1987, p.50).

Exposing the cultural dimension of artifacts – like the concept of a plan which is currently used in cognitive science – is a vital contribution to helping us to see the limitations of the metaphor of knowledge as mere information processing, which still considers computer systems as the best model of human cognitive functioning. The approach of situated action underlines the actor–environment interaction by assuming that action-in-context cannot be fully captured by any preconceived cognitive schema.

CHAPTER THREE

Decision-making, Appropriateness and Identity

One of the primary ways in which individuals and organizations develop goals is by interpreting the actions they take, and one feature of good action is that it leads to the development of new preferences. (March, 1991)

Intuitive, automatic decision-making, and even some deliberative decision-making, relies upon a simple comparison of each alternative of interest with a limited set of relevant standards, called images. On the one hand, these images are the decision-maker's imperatives, his or her values, morals, and beliefs – collectively called principles. On the other hand, these images are the decision-maker's existing agenda of goals and plans. If a decision alternative is incompatible with these images, it is rejected. If it is not incompatible, it is accepted. (Mitchell and Beach, 1990)

3.1 DECISION-MAKING AND CLARIFICATION OF PREFERENCES

The concept of a plan as an order capable of controlling daily action is in itself an artifact which allows some branches of cognitive science to preserve and promote the idea that rationality means the power of control that the mind has on the world. However, other lines in cognitive science have a more flexible concept of rationality. Johnson-Laird (1983, 1993), Johnson-Laird and Byrne (1993), Bara (1995) and Legrenzi *et al.* (1993) have all developed a cognitive perspective based on mental models understood as structures of analogic representations, highly sensitive to both contexts and situations. In their perspective, mental models reproduce the most relevant aspects of a situation and activate cognitive processes which depend on the particular configuration of the model. They consider thought as functioning not through abstract domain-independent procedures but on models specific for each area of application. Cognition, like action, is thus always situated in its context. Even in the case of deductive reasoning – an operation in which the idea of a universal mental logic appears to be plausible – the validity of this approach is affirmed; Johnson-Laird and Byrne (1993) state that rationality becomes problematic when it is presumed to be based on rules.

Clinging to rules in order to secure the intelligibility of cognitive processes is required more by our peculiar cultural context – which is now beginning to change – than by the need to provide a sound framework for the scientific understanding of problems. In the past, the scientific discourse has imported from the social context of Western cultures particular emphasis on rules and order, partly as protection against social and intellectual anarchy. But we are now beginning to achieve acceptable coordination in organizations, based on distributed decision-making instead of on control exerted by hierarchy. Thus, we now see that order is a concept that is wider than that of hierarchy (March, 1991) and that flexible order may serve us better in coping with unpredictable situations. Many scientific disciplines are now equipping themselves to face the unpredictable, discontinuity and disorder, both by means of new epistemological positions and by the chaos and catastrophe theories which thrive in the fields of physics and mathematics.

Our categories are now changing and are becoming more social. Distributed artificial intelligence (Gasser, 1991) aims at modelling in the language of artificial intelligence the characteristics of real-life environments in which it is normal for many people to participate in the interactions. In studies on decision-making – another fundamental activity mediating between action and situation – we also see a significant shift away from formal, syntactic models towards contextual, semantic ones. March, a Nobel prizewinner (with Simon) and an eminent professor at Stanford University, stresses the 'artifactual' nature of current theories on decision-making (1991, p.95). He considers decision-making (both theory and practice) as a ritual activity, closely connected with Western ideas of rationality. Our way of thinking of decision-making is a cultural artifact, just like the European – and Trukese – concepts of navigation.

Decision-making as a Western cultural artifact projects the idea of rationality as based on the gathering and processing of as much information as possible. In this perspective, the 'quality' of the decision depends on the greater or lesser completeness of available information and on the decision-maker's greater or lesser capacity for analyzing it. As we have already seen, thinking of situations as capable of being captured once and for all in a complete and 'objective' way is not only a dangerous illusion; it also distorts the very nature of the interaction between social actors and situations. Environments and situations are not 'data', they are constructed and reconstructed. Information is not simply collected, but must be sought after, according to certain aims which are appropriately assembled and interpreted. What is important for a competent decision-maker is not so much the *quantity* of available information – which under certain conditions may simply complicate matters – as its *quality*. Good information responds to the 'right' questions, i.e. it supplies pertinent responses to actors' primary goals.

The ability to manage information substantially consists of asking the right questions at the right time and place, and in interpreting the answers correctly. This is why, in particularly confused situations, expert managers rely more on dependable confidential items of news, their own impressions and personal scenarios than on the reams of paper spewed out daily by their office computers. In our culture, in which competence in managing information has supplanted our birthright as the legitimate source of authority, the right to command claimed by managers in organizations is based on their presumed – and ritually exhibited – capability to gather and process information. This may explain the paradoxical and yet universal situation pictured by March (1991): 'Decision-makers and organizations (a) gather information but do not use it, (b) ask for more and ignore it, (c) make decisions first and look for relevant information afterwards, and (d) gather and process a great deal of information that has little or no relevance to decisions. The generality of such phenomena suggests that perhaps it is not the decision-makers who are inadequate, but our conceptions of the role of information in organizations' (p.112).

It is the concept of decision-making as a rational calculation based on optimal management of information that March, after many years of top-level scientific research, criticizes and refutes on the basis of his observations of the ambiguity in actors' preferences in given situations. He believes that we must begin to recognize the fact that social actors are seething with discrepant and changing goals. It must also be admitted that disorder in interests and goals does not seem to worry actors excessively. They realize that their preferences are often inconsistent and may change over time, so that it is often difficult or impossible to predict what their future preferences are going to be; nevertheless, both individuals and organizations do not generally worry over much about putting order and stability into their system of preferences and eliminating sometimes considerable inconsistencies. That is, a certain amount of disorder is accepted as an essential part of daily life. The instability and conflicts of the system of interests may even appear as resources to be safeguarded: 'Though human beings seek some consistency, they appear to see inconsistency as a normal and necessary aspect of the development and clarification of preferences' (March, 1991, p.100).

It is thanks precisely to the instability of actors' systems of interests that new interests and new goals can emerge and be accepted. And so we see that, after two months of lessons, only now has Richard discovered a passion for physics he didn't know he had, while Jean, who has followed the same course, feels that physics is just not for her. The emergence of new preferences allows actors to grasp the opportunities the environment offers without jeopardizing their capacity to formulate long-term plans and stick to them.

Classical organization theories sought to give credence to the idea that, although individuals can give way to irrational behaviour, organizations are protected against this danger because of their impersonal character and hierarchical structure. Instead, recent theories, which consider organizations as constructions and networks of communication and cooperation (Jirotka *et al.*, 1992), see organizations as environments of distributed decision-making which are exposed to pressures very similar to those that afflict individuals. The result is that organizations can no longer be viewed as a kind of 'last bastion' of decision-making in terms of dispassionate rational calculation. Decision-making in organizations involves an ecosystem of actors searching to develop 'rational' strategies on the basis of limited knowledge, trying to identify their own priority goals inside a somewhat inconsistent system of preferences, and to assign plausible meaning to a confusing and ambiguous external environment.

3.2   DECISION-MAKING AS INTERPRETATION OF ACTION

In both organizations and individuals, decision-making is essentially a process of interpretation in which an active filter operates, very similar to that described in Chapter 1 regarding the actor–environment relation. On the one hand, this filter allows actors to select and interpret the relevant aspects of the situation; on the other, it allows the process of interpretation–assessment which identifies opportunities afforded by the environment to take place. The latter phase retroacts on actors' initial goals, which were used to construct and interpret the initial situation in a certain way, closing the cycle of goal–situation interaction with a new interpretation of the situation which is now modified by action. It starts up a new cycle of action–assessment according to new goals, new situations, new actions, and so on.

Figure 2 presents a diagram of the structure and function of the filter which links not only actor and environment, as we saw in Figure 1, but also situation, action and decision-making. Figure 2 depicts the main moments of the process of which situated decision-making consists: it starts with the selection of a certain course of action in response to a given situation (curve 1), continues with interpretation of the actual results of that action (curve 2), goes on to assess future prospects (curve 3), and finishes by retroacting on the interpretation of the new

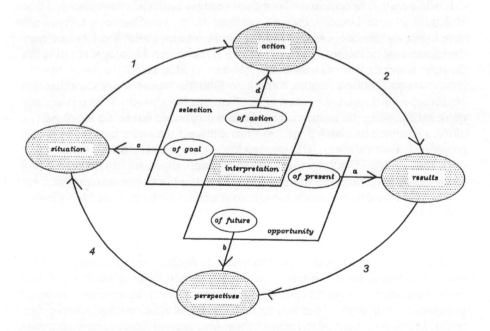

Figure 2 Decision-making evaluates actual results (a) and possible perspectives (b). Evaluation provides feedback about the interpretation of ongoing situations (c) and situated actions (d).

situation, which is formed according to the action taken at the beginning of the cycle (curve 4). Then a new cycle begins.

That making decisions means first of all interpreting accomplished action is given particular emphasis by March (1991): 'One of the primary ways in which individuals and organizations develop goals is by interpreting the actions they take, and one feature of good action is that it leads to the development of new preferences' (p.100). Clarification of preferences does not only provide criteria for establishing priorities among the requests made by the interest system to the mechanism of resource allocation available to actors; it also allows the development of goals in a qualitative sense, i.e. actors can discover new interests and goals through their actions: Richard, who until today had thought of nothing but the opposite sex, now discovers a passion for physics.

Actors adapt to situations first by selecting the actions carried out, and then by comparing expected and actual results. In both of these operations, the actors' activity is guided by their interpretation of the appropriateness of the action in question with respect to the potential offered by the situation and the meaning to be attributed to the result. In any situation, actors are not asked to decide initially on the intellectual level and then, only when they are ready, to translate their choice into action. They are obliged to do something at the moment the situation requires it, even if they feel they are not ready to make decisions or would prefer to stop and think first.

In daily reality, people often have to act not because they have decided that they want to, but because they are obliged to do so. The very fact of not responding, or postponing, is in itself a precise action: Philip has finished high school and now he has to choose what to do in the future. He may feel this is not the right moment to go to university, and if so in what faculty to enrol, but the situation requires him to act here and now. If he just ponders, does nothing and lets the university deadline pass without enrolling in any faculty, he has already made his choice by his inaction. The decision to choose not to act is not particularly appreciated in contemporary Western culture, although it is considered wise in others. In past moments in European history, the obsession about what to do was not so acute: for Victorian gentlemen, 'the time signature over existence was firmly *adagio*. The problem was not fitting in all the things one wanted but spinning out what one did to occupy the vast colonnades of leisure available' (Fowles, 1969, p.16).

If after high school Philip decides not to do anything for the time being, in reality he has already responded in a certain way to the question inherent in the fact that he now has a high school diploma. If Richard accidentally runs into Laura, the fascinating ex-girlfriend he is still in love with, he must do *something* by the very fact of having met her. Perhaps he has not yet decided what to do or what his reaction should be if he meets her by chance, but now he *has* met her by chance and he is obliged to face the situation, pleasant or unpleasant as it may be: should he scowl at her and turn away, without even bothering to say hello, or should he give her a beaming smile and try to get back on an affectionate footing? Whatever he does, he will later ask himself if he did the right thing, and will try to interpret *a posteriori* the effects and significance of his action and of Laura's response.

The idea that decision-making is an interpretation of what has been done is quite common in daily experience. We often read in the papers: 'General X and General Y met today, to discuss what happened yesterday at Z'. We have no difficulty in understanding that these are not cases of negotiation aimed at providing a joint and 'objective' estimate of losses and damage sustained as a result of the warlike action of yesterday. They are meetings in which two parties try to understand what has happened and compare their mutual intentions, evaluations and perspectives. This becomes necessary when the situation in question is particularly ambiguous; for example, there is a great difference in agreeing that the accident occurred by mistake instead of stating that it was a purposeful hostile attack, that it marked the end of a period of enmity instead of the opening of a new phase of warfare, or that what happened was an isolated occurrence instead of one aspect of a more general conflict.

Deciding means not just choosing between alternatives, but also interpreting the action to be undertaken, thereby capturing the emerging characteristics of the situation: 'Management requires tolerance of the idea that the meaning of yesterday's action will be discovered in the experiences and interpretations of today' (March, 1991, p.100).

3.3   IMAGES IN INTUITIVE AND AUTOMATIC DECISION-MAKING

In both individuals and organizations, the problem of decision-making in real-life contexts does not arise so much in terms of assessment of the means–end connection as in terms of identifying rules of appropriateness to be followed in interpreting situations and choosing proper lines of action. The rules are not right or wrong in themselves, but are more or less suitable for reconciling situational affordances with people's goals. In daily reality, decision-making takes place in the area of overlap between actor and environment (see Figure 1). The actor wonders: 'Have I given to the situation the response which best corresponds to what I am, wish to become, or want to communicate to others as being?'

In all cases, in a more or less direct way, intuitive decision-making involves the image of self. All voluntary actions form part of a plan, in turn at the service of a meta-plan occupying a certain position in the image of self which the actor constructs moment by moment by means of his situated decisions. So if Bill wants to study psychology at all costs, he does so because this meta-plan has some important relationship with the image of self which he is cultivating and wishes to project. In any given situation, when he has to decide on a course of action, the actor asks himself, 'What action must I take which is appropriate both to this situation and to myself?'

This question is double-ended, implying the selection of relevant aspects of the situation on the one hand, and an ordering of the actor's interests on the other. It is a situated problem, a choice between alternative lines of action here and now.

In general, for actor Y, studying physics while at the same time courting the lovely Z are not incompatible activities, but in this particular classroom and at this particular moment, he must choose between following the physics lesson or devoting his limited attention to Z's luminous smile and the intriguing things she told him just five minutes ago. Which of the two lines of action better corresponds to the project of self which Y is developing, and to his perception–assessment–appreciation of the elegance of differential equations as opposed to Z's charming smile?

When judging appropriateness, the possible match between the characteristics of the situation in which the actor operates and the salient characteristics of her or his self-image are both considered: 'Rather than evaluating alternatives in terms of the values of their consequences, rules of appropriateness match situations and identities' states March (1991, p.105). The fundamental question in appropriateness judgements is not, 'What good do I expect to get out of this decision?' but: 'Which choice is better – in the sense of suitable, appropriate, fitting – for a person like me in a situation like this?' It should be noted that actors move in situations with the aim not only of preserving their own identities and goals, but also of developing them; that is, discovering new preferences and thus having access to new environments, in a virtuous circle of enrichment of interests which coincides with the daily experience of those with a healthy sense of exploration.

The concept of decision-making as a bridge between situations and principles or social norms, recently suggested by the *image theory* of Beach (1990, 1993) and Mitchell and Beach (1990), is apt here. Image theory presumes that, inside their various situations, actors opt for certain lines of action with the aim of achieving goals which satisfy the principles, called images, which make up their identity. In this perspective, images are the elements which most closely control intuitive decision-making processes. Image theory offers an approach rather different from theories of decision-making as a rational calculation of advantages and disadvantages, benefits and costs consequent upon various possible courses of action.

The latter model of decision-making as a kind of double-entry bookkeeping, applicable when a judgement on only quantitative differences is required, finds its most widespread expression in the linear models of decision-making illustrated in the well-known letter that Benjamin Franklin wrote to Joseph Priestley in 1772 (quoted in Dawes, 1988, p.202): 'I cannot, for want of sufficient premises, advise *what* to determine, but if you please I tell you *how* ... My way is to divide half a sheet of paper by a line into two columns; writing over the one *Pro*, and over the other *Con*. Then, during three or four days' consideration, I put down under the different heads short hints of the different motives, that at different times occur to me for or against the measure. When I have thus got them all together in one view, I endeavour to estimate the respective weights to find at length where the balance lies ... I have found great advantage for this kind of equation, in what may be called moral or prudential algebra'.

Image theory starts by debating the utility of the double-entry kind of model suggested by Franklin and the validity of the underlying model of decision-making as calculation. In everyday reality, the theory notes that, first, very few decisions are taken according to comparative analysis of the pros and cons expected from every single possible choice; and, second, there are even fewer occasions on which actors resort explicitly to an estimate of the probabilities in question. It seems that the above two fundamental bases of traditional decision-making theory are not usually very evident in actors' ordinary experience (Mitchell and Beach, 1990). Traditional theories of decision-making as calculation and deliberation do not generally take into due consideration the processes taking place in real life, and formal analytical strategies are only occasionally used even by people who know and understand them. Moreover, even when individuals do use analytical strategies, they rarely accept the results if these clash in some significant way with their intuitions and general impressions.

When decision-makers make intuitive or automatic decisions, they do not really make cost–benefit calculations or probability estimates. They choose one alternative from a range of possibilities which are considered practicable and suitable. If this is the case, as image theory maintains, then there are two steps required to understand how intuitive decision-making happens: first, why is the range of admissible alternatives formed? and, second, how is one alternative preferred to the others?

In the image theory approach, the important moment is the first step, because it is here that the main selection is made. The mechanism which controls the admission of one option into the group of those selected for further consideration, and the final choice that is made in the second step, are critical for the formation of the decision, because they circumscribe drastically the range of available alternatives. The second mechanism, which is limited to choice of the best acceptable alternative among those that survived first examination, is clearly subordinate. This is even more evident if we reflect that, in some cases, only one alternative appears compatible with the decision-maker's acceptability criteria. In these cases, the first step also does the job of the second. Even clearer is the impact of the initial choice in those quite frequent cases in which the actor is unable to identify any acceptable alternative in the situation.

Mitchell and Beach (1990) believe that 'Intuitive, automatic decision-making, and even some deliberative decision-making, relies upon a simple comparison of each alternative of interest with a limited set of relevant standards, called images' (p.4). What are the images which act as ultimate standards when a decision-maker chooses an alternative? 'On the one hand, these images are the decision-maker's imperatives, his or her values, morals, and beliefs – collectively called principles. On the other hand, these images are the decision-maker's existing agenda of goals and plans. If a decision alternative is incompatible with these images, it is rejected. If it is not incompatible, it is accepted either as the final decision or as a member of the set of multiple alternatives that is passed on to a more complex mechanism that selects the best among them' (ibid.).

When Richard meets Laura in the street, he is still sensitive to the fatal attraction of the girl who left him and is therefore somewhat upset by their chance meeting and by the fact that he knows Laura now has another boyfriend. But he would never for an instant think of behaving badly or of insulting her, because such behaviour would be completely incompatible with his image of himself and of the situation. So if a friend of his, or a small devil seated on his shoulder, should hiss into his ear the suggestion that there is nothing to stop him insulting her if he wishes to, he would refuse without even thinking about it and say: 'Hey, I'm not that kind of person!' or 'She doesn't deserve *that*, surely!' In both cases, he would consider the possibility of behaving rudely or violently as repugnant, and his reaction would be something like, 'Who do you take me for? Do you really think I would do something like that?'

We see that Richard acts in this way because he has principles which are important to him: 'Implicit or explicit, central or peripheral, primary or secondary, the function of all principles is the same: they are the criteria for adopting or rejecting potential goals and plans. They are not the goals themselves, but they define what is and what is not desirable about goals; they are not plans themselves, but they define what are and what are not acceptable means for achieving goals' (Beach, 1990, p.25). These principles shape his personal and social identity, which Lazarus calls *ego-identity* (1991, p.101), and control the operations of the filter regulating the admissibility of options and courses of action.

3.4   APPROPRIATENESS, CULTURAL FRAMEWORK AND IDENTITY

In other situations, of course, things might not be so clear-cut for Richard. At a football match, he might easily feel like insulting the fans of the opposing team. He might seriously think of behaving in a way which he would consider wholly uncivilized outside a football ground; he might even go as far as to actually behave in that way. The problem is again that of selecting and interpreting the situations we started with: in what way does the chance meeting with Laura differ from the confrontation with the other team's fans? In which cultural framework are the two different episodes to be placed?

These questions may be answered by that part of image theory dealing with *decision-framing*. This consists of creating a scenario of the situation and the role of the actor in it, and is the phase just before the one we examined above, in which the set of alternatives which are admissible for the actor is formed. Decision-framing coincides substantially with our considerations in Chapter 1 regarding operations that take place in the space between the actor and his environment. Beach (1990) describes them in this way: 'We assume that the context has features, some of which derive from the knowledge the decision-maker brings to it and some of which are unique to the moment. These features are the cues the decision-maker uses to frame the context (p.51).

But situations, like interpretations, may change, which is one reason why actors may produce *misframes*. These occur either because the context has not been correctly identified, because an important piece of information has not been taken into consideration, or because the context has changed so that the preceding frame no longer fits it. The considerable number of errors and misunderstandings about which frame to fit situations into is of cultural origin. An environment may change its the cultural collocation without the actor being aware of it, appearing inside a scenario unlike the previous one in which it fitted neatly, and which the actor tends to reproduce in his framing of the situation without noticing that the context has changed. In such cases, automatic decision-making will result in a generally inappropriate outcome, because framing directly controls the automatic decision-making process. Instead, in intuitive decision-making, we may expect that the problem will not be recognized and as a consequence the range of alternatives considered to be acceptable responses to environmental demands could prove inadequate.

The cultural framework of decision-making is necessary for the decision-maker, enabling him to interpret events and himself in a way that is acceptably congruent with a repertory of meanings available both to him and to other actors. In this sense, the cultural framework is not in itself a product of the actor's relationship with the environment, but a precondition of his interaction and communication with it. Cultural rules exist before actors, which is why decisions regarding appropriateness generally occur almost automatically, i.e. they take place in a way that is not voluntary in the current sense of the word. They do not take into account – in terms of cost–benefit analysis – the future consequences that the present choice is expected to have on actors' current interests, but rather attempt to provide a reasonable fit between a changing, ambiguous set of contingent rules and an equally changing and often equally ambiguous set of situations.

However, in another sense, cultural rules do depend on actors, inasmuch as they are performed by them. They are embodied at the moment when the actors make them theirs, and incorporate them in their daily actions and choices. It is only at this moment that cultural rules become alive, since they have no autonomous, disincarnate existence in the heaven of platonic ideas. On the one hand, actors find themselves faced with cultural rules that act like an already existing repertory of meanings which stage characters can wear and a set of possible scenarios in which they can live. On the other hand, by absorbing them into their courses of action, they can transform and subvert them. But in all cases, social actors cannot help interpreting pre-existing cultural rules in the specific context in which they are involved, and they therefore inevitably modify them. Even the most inflexible traditionalist makes some subtle changes to the tradition he asserts he wants to maintain intact, when he presents it – incorrectly – as identical to its original in a context which is in fact different from the original one from which it drew its primitive original meaning.

The two poles – reproduction and cultural modification – both intervene in the formation of the self. Although every individual is somehow free to determine the relevant features of his own personal and social identity, he is always obliged to construct it with materials made available by the particular sociocultural context in which he lives, using models that are now mainly supplied by the media. The autonomy of the individual is both exalted and circumscribed by the symbolic order in which he acts. His hopes and fears, crystallized in the 'possible selves' he cultivates inside himself (Markus and Nurius, 1986), reveal the inventive nature of the self and at the same time reflect the extent to which that self is socially determined. The more an actor is competent in controlling the principles and meanings which his culture make available to him, the more he can adapt them to suit him better or even change them voluntarily; for example, he can understand the artifacts of his own environment and use them for his own ends, instead of being used by them (Mantovani, 1991, 1994a).

Social psychology has developed theoretical constructs covering the space between bottom-up processes linked to the construction of identity, and top-down processes linked to the transmission and implementation of cultural rules. We refer in particular to social identity theory (Tajfel, 1972) and to studies on social categorization processes (Abrams, 1992; Abrams and Hogg, 1990; Hogg and McGarty, 1990). Social identity theory defines social identity as the awareness an individual has of the fact of belonging to certain groups, linked to certain emotional values and certain meanings which are attributed to those groups. According to the theory, an individual tends to remain a member of one group, or attempts to become a member of a new one, when it seems to enhance social identity, increasing self-esteem and the satisfaction which is part of it. By means of categorization – the basic cultural character of which has been convincingly claimed by Lakoff (1987) – and identification processes, the gap between the *micro* level (that of actors' individual choices in daily experience) and the *macro* level (the meanings, values, what is appreciated and desirable in the framework of a given cultural environment) can be bridged.

Categorization and identity construction processes allow people to fit their own goals into the existing symbolic order, by physically or ideally belonging to groups which embody values they deem to be positive. This is why the group with which people identify themselves is only occasionally a group of people who are in physical reciprocal relation. In Tajfel's view (1972), belonging to a group indicates the result of a series of cognitive processes of categorization and identification, which must be distinguished from the meaning of the word 'group' intended to describe face-to-face relations among several social actors.

We usually use the word 'group' to refer to a link between individuals who interact directly, in physical mutual co-presence, like people who meet one evening at a party, or men and women working in the same office. But for Tajfel the word defines a connection between people who may not interact physically and directly, like 'the Dutch', 'red-headed girls', 'shy girls', or even 'fourteenth-century Florentine architects'. In Tajfel's sense, 'groups' do not so much define interpersonal direct relationships as social categories embodying certain values, which may be very large or quite vaguely defined: 'young female psychologists', 'fascinating women like me', or simply 'Seattle girls'. The important thing is that they should contain and project values to which individuals refer when constructing their own identity.

Inside the cultural repertory available to them, people choose values which act as orientative principles in defining their long-term goals and in constructing their identities (Higgins, 1990; Markus and Cross, 1990). Values are taken on not in their abstract and general formulations, but as embodied in the social realities with which people identify themselves. The term 'values' has no moral connotation here, but simply indicates what appears to be desirable to each actor: beautiful, true, useful, exciting, and so on. If Anne wants to be a psychologist, this is because she finds not only 'noble' or 'right' qualities in such a professional figure, but also pleasant and stimulating ones.

Social identity is not simply one part of a person's total identity, like a piece in a jigsaw puzzle. Identity is unique: it is not subdivided into two separate sectors, one personal and idiosyncratic, and the other social. Images of the self must be thought of as ranged along a continuum in which what makes one different from another are their specific contents, their complexity, the moment and the ways in which they are formed (Abrams and Hogg, 1990). Psychological research into social identity and categorization processes, illustrating how the link between social actors and cultural order is forged, makes an important contribution to our discourse. It tells us that individuals are connected to social structures by means of their self-definitions as members of various socially recognizable categories.

PART TWO

How a Shared Meaning is Developed

CHAPTER FOUR

The Social Context as Symbolic Order and Social Norms

A culture is that subset of possible or available meanings, which by virtue of enculturation – informal or formal, implicit or explicit, unintended or intended – has so given shape to the psychological processes of individuals in a society that those meanings have become, for those individuals, indistinguishable from experience itself. (Shweder and Sullivan, 1993)

People organize their projects and give significance to their objects from the existing understanding of the cultural order. To that extent, the culture is historically reproduced in action ... An event is a unique actualization of a general phenomenon, a contingent realization of the cultural pattern. (Sahlins, 1985)

4.1   SYMBOLIC ORDER, RECIPROCITY AND APPROPRIATENESS

Decision-making allows actors to develop and clarify their preferences by interpreting accomplished action and evaluating its results. In this sense, decision-making is a special type of action, symbolic. It attributes a definite sense to the match between situations and actions, thus offering an escape from or at least a substantial reduction of the ambiguity inherent in the match: 'Decision making is an arena for symbolic action and for developing and enjoying an interpretation of life and one's position in it. The rituals of choice tie routine events to beliefs about the nature of things. They give meaning'. (March, 1991, p.110). The meanings at stake in decision-making may be both general: actors take up positions on issues which are important for a large number of people, e.g. give their opinions on the political party they would like to see in power in their country; and local: actors take on commitments regarding the aspirations and needs of special individuals or groups, e.g. Bob chooses to study engineering, or to go skiing with his friends next weekend.

Interpretation is at the very centre of the three preceding chapters: constructing situations, situated action, and intuitive decision-making. Interpretation controls the filter mediating between social actors and their environment. In order to function, interpretation presupposes the existence of an order capable of assigning definite sense – at least partially shared among the social actors in question – to actions and situations. Cultural order is a framework that allows actors to attribute (at least tentatively) mutually recognizable meanings to situations and actions. As such, it is the premise which enables society, both entire and in its separate parts, to function in an acceptably predictable fashion.

When Marianne, the passionate heroine of *Sense and Sensibility*, writes in outrage to her deceitful suitor, she speaks the very language of appropriateness and reciprocity: 'I have passed a wretched night in endeavouring to excuse a conduct which can scarcely be called less than insulting; but though I have not been able to form any reasonable apology for your behaviour, I am perfectly ready to hear your justification of it'. She further clarifies her claim by writing: 'You have perhaps been misinformed, or purposely deceived, in something concerning me, which may have lowered me in your opinion. Tell me what it is, explain the grounds on which you acted, and I shall be satisfied, in being able to satisfy you. It would grieve me indeed to be obliged to think ill of you; but if I am to do it, if I am to learn that you are not what we have hitherto believed you, that your regard for us all was insincere, that your behaviour to me was intended only to deceive, let it be told as soon as possible' (Austen, 1986, p.199; original edition 1811).

Cultural order, to which all actors are bound to resort when looking for a repertory of possible meanings for actions and situations, establishes what is appropriate in a given social context, and also defines reciprocal obligations among actors. These two functions are not just connected but directly grafted into each other. Marianne requires the meaning of actions to be exhibited by the other actor and by herself, so that proper relations may be re-established between them. The attribution of meaning to actions accomplished or undergone (what she terms 'justification') and recognition of good faith and of the consequent reciprocal moral acceptance ('satisfaction') cannot be separated from each other. If her suitor behaves badly, refusing responsibilities which descend from his acts and implicitly assumed commitments, he is not a gentleman, and therefore honourable transactions with him are neither possible nor desirable. But if he *is* a gentleman, then his action must have some explanation which must be expressed in order for the relationship to continue and for the respective social roles to be confirmed.

Social and cultural norms institute roles, prescribing what is appropriate for actors in various situations and thus defining what each of them must do and what may legitimately be expected from others in different contexts. *Legitimate* is the key word here: Marianne may think, albeit with great displeasure and disappointment, that her suitor is abandoning her, but she simply cannot accept that such behaviour is appropriate for a gentleman in such circumstances. In Austen's world, if, after having made a commitment towards a young woman who reciprocates that commitment, a man suddenly disappears without notice, there is only one explanation: he is not obeying the rules one thought he would obey, because he is not the kind of person one thought he was. Instead, if that person *is* what we thought he was, he cannot have behaved like that. And in fact Marianne tells her family that she knows her young man well, he cannot have behaved dishonestly, and she still trusts him more than anyone else in the world. She cannot understand what is happening because her reconstruction of events and her expectations of her suitor come into conflict.

In the same way, a child does not doubt the role of his mother and her affection for him every time she takes him to the doctor for a vaccination. The local conflict between the meaning the child may attribute to the pain of the injection and his expectations of his mother as a source of protection is not resolved by questioning mother's role but by re-interpreting the situation. So when the doctor's nurse says, 'Now, I'm just going to prick your arm with this little needle and it may hurt a bit, but don't worry, it's good for you', we expect (or rather hope) that little Jimmy – even while possibly kicking and screaming – does not *really* believe that he has been abandoned by a monster-mother to the machinations of a cruel witch-nurse. The cultural framework may be sufficiently strong to compel actors to redefine the meaning of a given situation. If it is mother who says so, then going to the doctor for a vaccination is no longer interpreted as needless torture, but becomes necessary prevention and care. Even great pain may be redefined: according to context, we call it 'divine punishment', 'expiation of one's sins', or 'trials' to be overcome in order to accede to a higher level of awareness, as in the initiation rites of many cultures.

We saw, *à propos* of image theory, that Richard cannot even consider insulting Laura because such behaviour does not fit in with what he is and wishes to be. In that case, appropriateness was seen as a criterion for decision-making aimed at maintaining and developing personal identity. Here, it has a more general social role: it stabilizes social relations, ensures that the meaning of situations is shared, and justifies reciprocal expectations. Appropriateness rules set the stage on which actors move. And those actors interact easily because they know the script, and they believe that others know it too and will respect it. When the symbolic order on which a relationship is presumed to rest is absent or not respected, the consequences may sometimes be extremely unpleasant. Laura, who had originally gone out with Richard because she thought he was the nice young man he seemed to be, might discover that, once she had dropped him, he turned out to be a dangerous and violent bully.

What we are dealing with here is the reliability of relationships and the predictability of social contexts: only because she believes she is dealing with a young man of noble sentiments does Marianne venture to take on the complementary role of *inamorata*. Her personal identity and the choices it involves are set against a background of cultural norms and social rules: the appropriateness rules which define actors' intuitive decisions reflect the cultural norms which structure the social context. The result is that appropriateness is not only an internal question to be evaluated by one actor. It is basically an order which frames the actor–environment interaction. It may be disobeyed or transformed but it cannot be ignored, because it is the starting point for socially shared interpretation of the situation and of the action.

Cultural norms give form and scope to socially relevant actions in that they ground them in meaning as a fact, which is primarily public and only secondarily also private (Rosaldo, 1984). Shweder and Sullivan (1993) define culture as '... that subset of possible or available meanings, which by virtue of enculturation (informal or formal, implicit or explicit, unintended or intended) has so given shape to the psychological processes of individuals in a society that those meanings have become, for those individuals, indistinguishable from experience itself' (p.512). The area of overlap between actor and environment is occupied by interpretation, made possible by the presence of meanings as public facts. They structure our perception of the world on the one hand, and our situated action on the other.

The definition of what is appropriate in various circumstances is the basis of reciprocal obligations; on it the intelligible and predictable functioning of social context depends. The symbolic and normative dimension of the social world stands as the condition of reciprocity in actors' expectations, beliefs and behaviours, not vice versa. By respecting appropriateness rules, actors commit themselves to acting correctly in exchange for the implicit commitment to be in turn and at the right time treated in an appropriate way. Correctness is defined inside the system of expectations embodying various actors' roles: in a given situation, behaviour which is 'correct' for a doctor, for instance, may not be so for a boyfriend or the office boss.

The existence and persistence of rules, combined with their relative independence of idiosyncrasies, enables institutions (March and Olsen, 1989), organizations, groups, and individuals to function with an acceptable degree of reliability: people know what they can and must give and receive from, respectively, the office boss, the boyfriend, or the doctor, and their expectations are usually confirmed by experience, if the cultural system functions properly.

4.2   RITUALS AND THE DISCREPANCY BETWEEN VALUES AND EVERYDAY LIFE

The relationship between symbolic order and reciprocity in actors' conduct is often reconstructed in a distorted manner in the social sciences. Instead of recognizing that symbolic order generates reciprocity, it is presumed that reciprocity generates symbolic order. Those who adopt this point of view consider social rules and cultural order as simple consensual meta-games whose rules are more or less agreed upon by those who play them.

We must recognize that 'to some extent there certainly appear to be such implicit "contracts", but socialization into rules and their appropriateness is ordinarily not a case of wilful entering into an explicit contract. It is a set of understandings of the nature of things, of self-conceptions, and of images of proper behavior' (March, 1991, p.106). This inversion of priorities, which places rules as the *result* of negotiation among actors rather than its *premises*, pervades all contemporary social psychology. This point will be discussed later, especially in Chapter 7 on communication. For the time being, we stress that between the symbolic dimension and the interactive (interpersonal) dimension of the social world, the former precedes and makes possible the latter. It is the existence of meanings, in a certain sense anteceding interaction, which confers significance on the interaction between social actors and their environment. In turn, they influence the initial symbolic order, implementing and at the same time transforming it. Like Sahlins (1985), we could define symbolic order here as a *virtual structure*, and the action of the actors who implement it as an *actual structure*.

The priority of rules with respect to interactive behaviour is clear in the interpretation of ritual of Jonathan Smith (1982). He observes that, in the hunting rituals found in sometimes very different cultures – from the Yakut who hunt bears in Siberia to the Pygmies who hunt elephants in Africa – there is a set of elaborate rules of etiquette which must be strictly observed if the animal killed is to be considered acceptable as proper food by the tribe. The rules imposed by the ritual are generally rather difficult to apply. For instance, they may prescribe that the animal may only be struck while it turns its head towards the hunters or runs towards them, that wounds may only be inflicted on certain parts of the body, or that the hunters should make a short speech justifying their action and praising the animal before they kill it.

Clearly, in everyday life, tribes that depend essentially on hunting for their survival do not usually behave in the prescribed way during the hunt and cannot allow themselves to be too difficult to please in accepting as food meat from animals which have been killed in an incorrect or questionable way. Why, then, establish such restrictive rules, so far from the daily reality of hunting and eating? What meaning can be attributed to rituals which are presumably often not respected? If cultural rules were the product of direct interactions between actors and situations, we should find in them the reflection of current practices, not the opposite, as seems to be the case here, where we see that rituals do not codify accepted daily conduct but contest it.

Smith's explanation for this singular situation, in which ritual challenges daily reality, is that ritual has the function of detaching the daily world from the ideal one. The actions we accomplish in everyday life, compelled as we are by our contingent needs, in confused and often hazardous situations, rarely fit what we think and know we should do. Ritual recognizes and elaborates the discrepancy between symbolic order and everyday reality: 'It provides the means for demonstrating that we know what ought to have been done, what ought to have taken place ... Ritual provides an occasion for reflection and rationalization of the fact that what ought to have been done was not done, what ought to have taken place did not occur' (p.63). The very existence of ritual indicates that actors are aware of how things should be done in a perfectly ordered world: 'Ritual represents the creation of a controlled environment where the variables of ordinary life may be displaced precisely because they are felt to be so overwhelmingly present and powerful. Ritual is a means of performing the way things ought to be in conscious tension to the way things are, in such a way that their ritualized perfection is recollected in the ordinary, uncontrolled course of things' (*ibid.*). Ritual criticizes situations rather than reflecting them.

Rules are principles to interpret events, not their effects. They are criteria for judgement, not items in a balance sheet closed after the *fait accompli*. Ritual allows actors to be dissatisfied with what they actually do, even though they know that this is necessary for their survival. The Yakut hunter knows very well that, if he wants to kill his bear, he cannot strike an attitude in front of it and recite a speech exalting the reverence and awe the tribe has for its mythical protector or ancestor. He is also convinced – and ritual forces him to make public testimony of this – that one cannot kill with impunity, that the bear is sacred, and that, if these principles were not recognized and affirmed, life would lose much of its savour.

Ritual combines these discrepant pieces of knowledge without reconciling them. It juxtaposes but does not blend them. The contradiction between ritual and practice, between principles and daily life, enhances human experience, which is seen to fall between the two poles of what should be done and what is in fact done. Practices alone cannot offer a justification of themselves which goes beyond their transient success. In the tension between daily practices and cultural rules – which are the same values and principles of Beach (1990) and Mitchell and Beach (1990) – existence becomes meaningful and interesting. The forbidden love of Anna Karenina confirms and at the same time contradicts the rules which make that love both scandalous and irresistible.

In modern organizations too, ritual plays an essential role, because in modern, complex societies it is even more necessary than among the Yakut or Pygmy hunters to know which principles must be followed, which rules govern the game one would like to play. The contests that occur both inside organizations and among them select and clarify the principles which are to be adopted as the best suited to interpret situations. March (1991) believes that decision-making in organizations does not consist first and foremost in choosing between alternatives. It is more a matter of identifying which criteria are acceptable for interpreting situations: 'Life is not primarily a choice; it is interpretation. Outcomes are generally less significant – both behaviorally and ethically – than process. It is the process that gives meaning to life, and meaning is the core of life. The reason people involved in decision making devote so much time to symbols, myths, and rituals is that they care more about them' (p.111).

It is by means of ritual that actors seek to overcome the discrepancy between values and actions without giving up either of these two conflicting elements: the action which fits circumstances, or symbolic order which lays down what should be done. Executives aware of the real meaning of their function cultivate ritual not only to legitimize their decisions, but also and especially as an indicator of the persistence in daily life of the need for symbolic order which makes even inconsistent or opposing actions meaningful.

4.3   MISFRAMING: WHAT HAPPENS WHEN MEANINGS ARE NOT SHARED

We have seen that Marianne, Richard, and little Jimmy at the doctor's surgery all construct situations starting from mutual expectations and from the system of rules which justify them: 'An event is not simply a phenomenal happening, even though as a phenomenon it has reasons and forces of its own, apart from any given symbolic scheme. An event becomes such as it is interpreted. Only as it is appropriated in and through the cultural scheme does it acquire an historical significance' (Sahlins, 1985, p.xiv). Events are not raw data, but interpretations constructed according to a certain cultural order: this reflection further develops our theme by indicating that it is actors' interests which confer a definite physiognomy on the environment. Now we emphasize the fact that actors' interests come into being inside a pre-existing network of meanings which frames them.

Cultural order, although it does not restrict actors' freedom of movement in that it does not compel them to comply to norms, does in fact somehow constrain their autonomy in attributing socially recognizable meanings to situations and actions. Cultural structure controls events only partially because, as Sahlins notes, events keep their internal force, which hinders any attempt to capture them fully inside a predefined system of meanings. Symbolic order is a grid, or rather a set of interpretative grids, that is indispensable but not exhaustive. It is a map, not the territory itself.

It may happen that a certain portion of territory is poorly mapped. For example, in the Victorian world of Fowles (1969), the characters may fall not only in love but also into despairing confusion, because their map equates impetuous emotions with the loss of self-control. When Charles meets Sarah, the French lieutenant's woman, he is bewildered by the fact that the girl asks for his help and in the same breath contradicts him vehemently. This sort of thing was not permitted in the Victorian culture: 'A woman did not contradict a man's opinion when he was being serious unless it were in carefully measured terms. Sarah seemed almost to assume some sort of equality of intellect with him; and in precisely the circumstances where she should have been most deferential if she wished to encompass her end. He felt insulted, he felt ... he could not say' (*ibid.*, p.116). The embarrassment of Charles who, already at this point in the tale, begins to play the uncomfortable role of a young *man* who is seduced and abandoned, arises from a profound misunderstanding of the situation. As he has no idea of what seduction means, he is unable to say what he feels when confronted by Sarah, and cannot recognize in himself the symptoms of emotions which he has only read about in books. 'Perhaps he had too fixed an idea of what a siren looked like and the circumstances in which she appeared – long tresses, a chaste alabaster nudity, a mermaid's tail, matched by an Odysseus with a face acceptable in the best clubs. There were no Doric temples in the Undercliff; but here was a Calypso' (p.117). Charles, we see, does not possess a suitable reference grid that can help him to make sense of his new situation.

Similar confusion was experienced by Atahualpa in his first, catastrophic encounter with the Spanish *conquistadores*. He did not understand what was happening, not because he was not an intelligent man and a competent monarch – he was in fact a godlike figure in his own social and ritual environment – but because he did not have a symbolic framework which fitted the situation and which would have allowed him to understand the novelty inherent in the arrival of these uncouth foreigners. At the very moment when the dramatic encounter at Cajamarca was so crucial for Atahualpa and for his people, the Inca king found himself without certain critical competences, due to his cultural background, not to his individual capabilities or lack of them.

The more situations appear to be incomprehensible, the more actors need to find plausible and socially shared explanations for them. We generally try to give some meaning to present bewildering experiences by comparing them with similar happenings in the past, as the Victorians did when they tried to exorcize love by likening it to an episode of drunkenness or an illness that caused a fever which generally cleared up in a short time without after-effects. We orient ourselves in new situations by drawing on pre-existing cultural references and suiting them to situations which are often far from those foreseen in the original cultural map. In doing so, we often resort to metaphors, explicit or implicit. We use metaphors not only to communicate and negotiate the social meaning of events, but also and particularly to produce satisfying categorizations of situations (Lakoff, 1987) in which the cultural grid making them accessible is normally incorporated. Considerable degrees of misunderstanding may arise between actors when their symbolic contexts are too far removed from each other and at the same time events do not immediately reveal this fact.

We may not be aware that, in our sense-making activities, we are using different metaphors or the same metaphor in diverging acceptances, so that the meanings of reciprocal actions are not shared. There are circumstances in which we may not even realize that misunderstandings are arising. The problem with metaphors is that even those who produce and use them are not always aware when their creatures leave their control and stop being applicable to the intended situation; that is, when they no longer supply an acceptable map of the territory and cause us to lose our way.

Metaphors establish an analogy between an original, familiar domain, and a target domain, which is explained by reference to the former. If Bob says: 'My marriage is a prison', he presumes that his interlocutor is familiar with the idea of a prison and establishes a connection between the prison situation (the origin of the metaphor) and his marriage situation (the target). He uses the prison metaphor to try and make sense of what is happening to him, as well as to communicate his plight to his friends. But this metaphor has its limitations, which may not be completely clear even to Bob, with the result that his interlocutors may be even more confused. How far can this metaphor be applied? To what extent does the analogy between a gloomy dungeon and Bob's marriage hold good literally and to what extent only figuratively, and where does it stop being applicable at all?

Inside a widely shared symbolic order, like that offered by a certain culture to its members, the symbolic frame upholding metaphors is generally sufficiently robust to allow actors to make acceptably efficient categorizations and communications. But what happens when actors confront each other on the line of demarcation between two completely different worlds? How can they come to understand the reciprocal intentions, the relevant characteristics of respective social environments, and the current meanings of metaphors which regulate inter-actions? In order to understand the extent to which the cultural frame influences daily experience, we might refer to what can occur when the mediation of a shared cultural order is absent.

Let us recall, for example, what happened when Captain Cook's ships arrived in Hawaii, over two centuries ago. Sahlins (1985), professor of anthropology at the University of Chicago, comments: '7 December 1778. The *Resolution* and *Discovery* were beating against the wind off the north coast of Hawai's Island. On this day, Captain Cook finally relented and granted Hawaiian women the right to be loved that they had been demanding since the British first anchored at Kaua'i, January last, discovering these Sandwich Islands to the Western world'. Sahlins goes on: 'At Kauati, Cook had published orders prohibiting all intercourse with the local women, for fear of introducing the "Venereal Complaint". But the same pages of his journal that record these orders also convey Cook's sense of futility. Similar measures had already failed at Tonga, and the behavior of the Hawaiian women was even more scandalous' (p.1).

This picture of uninhibited eroticism combined with tropical exuberance inside the ordered, sophisticated Hawaiian culture may appear idyllic to modern readers, but it did present problems to European culture at that time. Cook's reports and to an even greater extent those of the French explorer de Bougainville were as avidly lapped up in Paris as they were in London. On the one hand, they launched the enduring myth of exotic Polynesia and, on the other, they criticized the bleak-ness of repressive European customs. Sahlins vividly depicts the scene beheld by Cook's adventurous sailors: 'By the time the *Resolution* reached the south coast of Hawaii, he [Captain Cook] was complaining of the difficulty of working the ship with so many women about. Saswell [surgeon's mate] had no complaints: just a wave of the hand, he said, could bring a "handsome girl" to the deck, "like another Venus just rising from the waves"; and when the British finally anchored at Kealakekua Bay, "there was hardly one of us that may not vie with the Grand Turk himself"' (*ibid.*, p.3).

Clearly, although the Polynesian women's costumes were deplorably scanty owing to the temporary lack of missionaries, Cook's men were unable to see the reality staring them in the face. So what *did* they see, what did they write about in their diaries? They saw Venus, a Greek goddess. They saw the European image of beauty and love; a familiar image, all things considered. They had their cogni-tive schemes – their categories – on which to model and assimilate their new travel experiences, although these were very disconcerting at first sight. Actually, a little thought was enough to show that these Hawaiian beauties were neither blond nor blue-eyed, nor did they move, laugh or sing like European girls. And their dancing was something entirely different from Greek or British dances.

So another metaphor emerged, which allowed the surgeon Samwell and the sailors to make sense of the situation: 'If we are in a place where Venus exists without the restrictive British customs, we might as well imagine being in Turkey. Even better, each of us is a sultan, the Grand Turk, surrounded by dozens of beau-tiful women offering themselves as if such a thing were normal ...'. As we know, the European imagination had done a great deal to embroider the erotic, trans-gressive resources of the Levant and of the harem in particular. Metaphor again allowed the sailors not to feel too lost: 'We needn't worry if we are far from our own world. We are not lost in a parallel universe. We haven't strayed too far, we are in Turkey or some crazy foreign place like that. And things are wonderful, we're sultans!'

Metaphor is more than a paragon or a cognitive category. It helps us cope with new situations, supplies us with a pragmatic criterion, serves as an emotional stabi-lizer. It was all very well for Cook's sailors, thanks to their ideas about Venus and the Grand Turk. But how did the Hawaiian women view the situation, not to speak of their esteemed husbands, brothers and fathers? The women of the South Seas had dances, songs and initiation rites which would have been considered wicked and in any case totally unsuitable for social occasions in Europe. Venus herself might have been a little perplexed if she had been requested to take part in them. Her classic static pose was a far cry from the Hawaiians' energetic, hip-wiggling, sacred dance, the *hula*, which had the aim of making the new year fruitful by sexu-ally arousing Lono, the god of cosmic reproduction.

The sexual practices of the Hawaiian maidens aboard Cook's ships were, Sahlins explains, an essential part of the sacred annual celebrations in honour of Lono. The European sailors clearly did not have the faintest idea that they were participating in holy rites, and would probably have been amazed to learn that their couplings had any meaning different from the (to them) obvious one imposed by their original European cultural context. Sahlins stresses the equiv-ocation into which the Europeans had fallen in interpreting the women's behaviour: 'The women offered themselves because they thought there was a god, and the sailors took them because they had forgotten it' (*ibid.*, p.5). The apparent initial harmony and equally apparent enthusiastic cooperation between Euro-peans and natives, actors from two very distant cultures meeting for the first time, hid a profound cultural divide which was to be revealed only later.

The Hawaiian men too had cast their spells; for them too the contact with the English was the fulfilment of centuries-old rites. They too were moving according to a symbolic order, but theirs was far removed from that of their European guests. And it was not even true that no taboos had been broken. Many *had* been broken, into small pieces. But they were not the taboos the Europeans had expected. Native taboos did not prohibit sexual activity, which on the contrary was highly regarded and indeed encouraged, but the eating of certain foods by women and by men in the presence of women. 'On board the ships, the sailors were drawing Hawaiian women into their own conceptions of domestic tranquillity. They invited their lovers to sup, on such foods as pork, bananas, and coconuts. So did the women doubly violate the strictest Hawaiian tabus on intersexual dining. Custom-arily, men's meals were taken in communion with ancestral gods, and these very foods were the sacrifice, hence at all times prohibited to women. The participa-tion of the women would defile the sacrificer, the offering, and, for that matter, the god' (*ibid.*, p.8).

Since they ate with the Hawaiian women, the Europeans were soon defiled without knowing it. The apparently obvious initial bond between the two peoples rapidly disintegrated. Mutual incomprehension manifested itself and grew, culminating in a series of misunderstandings in which the actors of one culture were unable to make sense of the actions of the other, until Cook, ritually identified with Lono, was killed. The Hawaiians had acted throughout according to a cultural scenario of which the European explorers were completely ignorant and of which they did not suspect the existence, even after the bloody epilogue of Cook's murder. Both parties had constructed an explanation which did not cover the facts: a dramatic example of *misframing*. A series of misunderstandings in interpersonal and intergroup relations was followed by a true conflict (*quarrel*), leading to a physical confrontation with lethal consequences to both sides (*battle*). The escalation of the conflict until it exploded in physical violence was only the translation into proper conduct (and, as such, binding on all the decent, self-respecting members of both cultures) of the different interpretations which the two cultures gave of the ongoing situation.

Symbolic order, if it is not shared, cannot ensure the intelligibility and reciprocity of actors' conduct. Conversely, when symbolic order not only functions properly but is also enhanced in its role as map of social and physical environments, communication and cooperation among actors are greatly facilitated (Mantovani, 1996).

4.4 A CONCEPTUAL MODEL OF THE SOCIAL CONTEXT

We now see that the construction of contexts is not a free invention, in which actors arbitrarily assign meanings to the various environmental configurations which confront them, but an activity to which a pre-existing set of rules and meanings is applied. Human beings organize their self-images, identify their values and formulate their plans starting from the attributions of possible meanings according to the premises inherent in the existing cultural order. In this sense, writes Sahlins (1985), we may say that 'the culture is historically reproduced in action ... an event is a unique actualization of a general phenomenon, a contingent realization of the cultural pattern' (p.vii).

Conversely, when acting inside the symbolic order which modulates their daily experience, actors cannot help transforming that order. They can change the meanings attributed to given actions or situations; alternatively, the circumstances may be novel and as such may elude appropriate traditional categorization: '... As the contingent circumstances of action need not conform to the significance some group might assign them, people are known to creatively reconsider their conventional schemes. And to that extent, the culture is historically altered in action' (*ibid.*).

In reality, situations are always to some extent novel, if we look at them carefully enough, closely enough, and above all if we ask the right questions. If we notice how innovation insinuates itself silently and inexorably in the cultural order every time innovative conduct, an unknown artifact, or an as yet uncoded form of interaction emerges, we can see how cultural order as a whole is constantly being transformed by current action. Every significant cultural change modifies the overall structure, since changing some meanings alters the relationships interacting between all the other categories: in this sense, symbolic order is intrinsically changeable, it is an 'historical object' (*ibid.*).

The main source of alteration and change in daily situations lies in the mutation of the cultural frame in which they are located. Situations cannot be interpreted directly, outside any cultural map. Although contexts are 'resources for interaction created by participants' (Cooper, 1991, p.243), they are not produced simply according to arbitrary preferences. It is this pivotal sense of the conceptualization of social context that we outline below (Figure 3).

This model has three interdependent levels. The *first* level is that of the social context in general, the *second* that of daily situations, and the *third* that of local interaction with the environment by means of artifacts. The three levels nest inside each other from bottom to top. The use of artifacts is a particular aspect of daily situations which, in turn, is included in the more general social context. From the top level downwards, we have the key to the interpretation of the lower levels: the social context supplies the elements which allow situations to be interpreted. Situations in turn inspire the goals orienting local actor–environment interaction which takes place through artifact use.

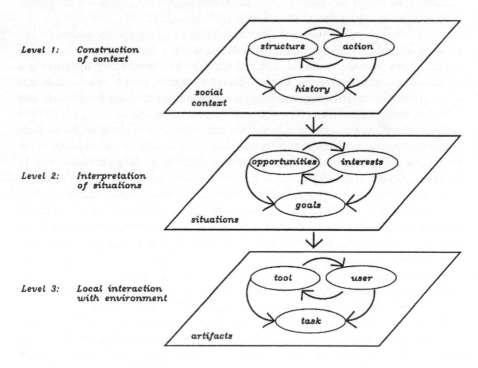

Figure 3. Three-level model of social context.

At the highest level, social contexts are conceived as constructions which are produced and continually reproduced by the encounter of situated action with symbolic order. Change – the history of the context – is the result of this encounter. Let us comment briefly on these passages:

(a) *Contexts*: These are not data, but are built, which is why the social context eludes attempts at capturing it by means of a descriptive approach. The constructive feature of the context supplies a justification of principle, and not only of fact, of the existence of irreconcilably different viewpoints among actors. This is not one of the many flaws of human beings; the plurality of perspectives on situations arises from the fact that the context is constructed in ever-divergent ways due to the effect of the alteration which action produces on the cultural structure in which it occurs.

(b) *Structure*: This is the first element composing context. It designates the set of symbolic relationships making up a culture (Archer, 1988). It is also called here 'symbolic order' or 'social context', in its widest sense. It constitutes the repertory of images or principles which Beach (1990) places at the base of the decision-making process. Each individual then adds to, removes from, or modifies elements of the repertory, so as to obtain that 'similarity without identity ... allowing us to be individuals instead of cultural clones' (p.24). It is the condition of cooperation and communication: 'To function successfully within a culture, be it only two people, as in the marriage, it is necessary for the participants to share principles' (*ibid.*).

(c) *Situated action*: This is the second element making up the social context. Situated action, presented and discussed in the preceding chapters, assumes that people use situations as points of departure to generate goals and plans suitable for specific circumstances. It therefore refutes the idea that action may be isolated from situations and represented as the execution of prefixed plans constraining the action.

(d) *Structure–action relationship*: This is composed of two phases. In the first, the structure influences possible socially significant actions within itself. In the second, actors constantly re-elaborate traditional models in a creative way. In this sense, culture is transformed by action and by actors' choices. The result of this two-phase process is that the context is intrinsically unstable.

(e) *History*: Changes in the social context are not casual, but are the inevitable consequence of human action. When implementing practical plans and social relations, we reformulate cultural categories by means of our daily practices.

In this model, the social context is essentially the structure of a certain culture, constantly subjected to alteration by practical human intervention. It cannot be reduced simply to interpersonal relations, understood as the (possibly physical) interaction environment in which an information exchange takes place. The social context is the precondition of communication: a shared symbolic order in which action becomes significant. Holyoak and Thagard (1995, p.211) illustrate how, through analogy, the cultural order is constantly able to influence everyday people interactions:

> The web of culture that holds people together in social groups is constructed from shared beliefs and feelings, knowledge of a common history, and a sense of place in the natural and social world. These strands provide the connections by which members of a society can communicate with one another. Myth and magic, rites and ceremonies, poetry and everyday conversation all form part of the web. A culture is built and maintained in large part by symbolic stories and rituals, in which objects and events are given meanings that in various ways go beyond themselves. Analogy plays a prominent role in providing these extended meanings and thus in building and maintaining the web of culture.

Social actors do in fact exchange meanings, not little pieces of information. More precisely, they exchange interpretations of the situations in which they are involved. In this sense, the key to the clarification of their messages is their reference to an at least partially shared meaningful context.

This socio-cultural approach is beginning to be appreciated in social and psychological studies, which previously focused on individual processes. We realize that we cannot create an unnecessary and misleading boundary by separating individuals from their culture. In real life, individuals do not stand alone against their socio-cultural environment; they are profoundly inserted in it. So pervasively do cultural systems regulate people's everyday activities (Rogoff *et al.*, 1993) that it is incorrect to say that we 'acquire' a culture (Bruner, 1993); what really happens – Bruner says – is that we 'enter' it, or 'are enabled' by it to live in it: 'Culture is not a set of responses to be mastered, but a way of knowing, of constructing the world and others' (p.516).

Until now, we have described the top-down path of our model. We have shown how meaning descends from symbolic order within situations – and from them within local interactions with artifacts – by means of interpretative processes. However, we must consider (in Chapter 5) the fact that there is also a bottom-up path in our model, a *pragmatic* approach to cultural order. Actors influence the physical and social environment by their actions and initiate a spiralling exchange with the environment which starts from local interation with artifacts, rises to modify daily situations, and ends by influencing symbolic order as a whole. Artifacts are not inert, asocial objects. Is there anyone who seriously doubts that telephones, cars, or television have no cultural importance?

CHAPTER FIVE

Technological and Normative Artifacts as Embodied Projects

▉

If one believes that people 'think through artifacts' and that the shape of this thinking is constrained by the way the particular set of artifacts is put together as part of a historical stream of human activity, the motives for which are survival and reproduction, then experimental procedures that grew out of late 19th-century science cease to be plausible as generally appropriate to the study of human thinking. (Cole, 1990)

The material alterations in their means of production were manifested in transformations at intimate levels of experience – assumptions about knowledge and power, their beliefs about work and the meaning they derived from it, the content and rhythm of their social exchanges, and the ordinary mental and physical disciplines to which they accommodated in their daily lives. I saw that a world of sensibilities and expectations was being irretrievably displaced by a new world, one I did not yet understand. (Zuboff, 1988)

5.1   PRACTICES AND INSTRUMENTS: THE TASK–ARTIFACT CYCLE

Events and situations are necessarily mapped in symbolic order, but the resulting map does not exhaust their potential meanings, since there is always space available for pragmatic reformulation: 'In their practical projects and social arrangements, informed by the received meanings of persons and things, people submit cultural categories to empirical risks. To the extent that the symbolic is thus the pragmatic, the system is a synthesis in time of reproduction and variation' (Sahlins, 1985, p.ix). In the interpretation of each of the innumerable aspects of daily situations, precoded top-down meanings deriving from cultural tradition and new bottom-up meanings generated by practices meet and confront each other. The bottom-up meanings arise from everyday interactions between actors and situations as well as from the situated use of artifacts, and need to belong to symbolic order if they are to become meaningful.

On the one hand, we have the priority of the existing cultural order, of tradition, from which innovation itself depends, in order not only to be understood and communicated but also simply to exist as a socially recognizable event. Empirical facts are not accessible as such, but inasmuch as they receive appropriate sense from the culture in which they appear, to be processed and interpreted as being good or bad, worthy or unworthy of attention or support, and so on. Thus, the old cultural system is projected forwards into the present and future, investing new events with its meanings too. Every innovation must be grafted on to preceding tradition, as is happening as the new technologies of information and communications spread. Society uses technologies, not vice versa. Even when technologies produce a break, an indisputable discontinuity, within their social and organizational contexts, the latter always manage to use new technologies for their games and for pre-existing goals (Pinch and Bijker, 1993), social continuity prevailing over possible technological discontinuity. This is why the social sciences find it plausible to privilege traditional and institutional forms rather than sets of innovative practices, thus following a one-way path along which actors' conduct is reduced to applying the rules of the pre-existing situation.

On the other hand, however, situated action continually and inevitably reformulates the entire symbolic order starting from everyday practices. Through practice, actors make real that cultural order which, if it were simply inherited from the past like the Hammurabi Codices, would only remain a repertory of abstract possibilities. Principles do not inhabit a world apart, but are produced and reproduced moment by moment in the interaction between actor and environment: 'Friendship engenders material aid: the relationship normally (as normatively) prescribes an appropriate mode of interaction. Yet if friends make gifts, gifts make friends ... The cultural form can be produced the other way round: the act creating an appropriate relation, performatively' (Sahlins, 1985, p.xi).

Symbolic order is not what is engraved on the tablets of law, but what functions as sense-making and norms in daily interactions. Of course, if the engraved tablets have some credit inside a given social context, they will directly influence actions, but not necessarily in the sense that they conform almost automatically to rules and principles. Rather, norms often act in the same way that ritual influences the hunting practices of the Siberian Yakut: they point out what is decent and just in a given situation, even if and especially when actors' behaviour diverges from decency and justice. An examination of practices thus offers an essential and complementary contribution with respect to that provided by consideration of the cultural order.

Situations are illuminated on two fronts: top-down, from the cultural tradition of a given social environment; and bottom-up, from the real interactions between actors and environments expressed in daily practices and in the effective use of artifacts. The negotiation of the meaning to attribute to situations consists both of integrating practical innovations in the normative social context, and of adapting that context to the continual changes produced by individual and social practices. The first task allows us to understand innovation in the light of traditional categories; the latter transforms the existing order, enfolding into itself the meanings emerging from the new practices.

Practices, understood as local interactions in which new interpretations are sometimes produced, may be categorized, at the lowest level, as the performance of tasks in view of certain goals. Artifacts are normally present in task perform-ance, in work and in free time. Artifacts mediate between the actor and the envi-ronment, they weave in and out and mingle throughout social and physical environments. This is why Woods and Roth (1988) say that environments cannot be violently stripped of the tools which form part of them, and recommend the development of 'an ecology of tasks and artifacts'. The use of instruments and the ways of performing tasks influence each other reciprocally. Carroll and Camp-bell (1989) describe the task–artifact cycle thus: people perform certain tasks, with more or less satisfaction, using certain tools. These tasks are the starting point for the invention of new artifacts which aim at getting people to perform their tasks even better, within the bounds of available technological capacity. Once adopted, the new instruments alter the tasks for which they were originally designed, and modify the situations in which those tasks were previously performed.

This gives rise to the need for further innovations, made possible thanks to a new generation of artifacts to be invented in order to perform the new task even better, and so on, *ad infinitum*. This circular process prompts the incessant tech-nological innovation which is so characteristic of modern Western culture, which considers the environments of human activity as spaces within which infinitely improvable tasks can be performed by applying ever-improving instruments. Tech-nological development embodies an almost inexhaustible potential for cultural transformation which coincides with its very nature. If technology aims at producing increasingly more efficient instruments and if, by so doing, it contin-ually transforms the physical, social and cultural environment which contains it, then it stands as a factor that permanently subverts the traditional order. Modern Western cultures, which reserve a place of honour in their system of values for technological innovation on the one hand, and political democracy on the other, cultivate elements which oblige them to a continual, wearying reformulation of their traditional repertory of principles and values.

It would be misleading, however, to think of technologies as separate from actors (Hakken, 1993). The task–artifact cycle originates from the interests of actors which oblige them to do certain things. A technological tool does not tell people or organizations what they want to do or must do. In real-life conditions, technology and culture are intimately entwined, but if we were to try to separate them artificially culture would have the last word. Technologies are produced and used by social actors for their goals; they are not independent realities. When tech-nologies are successful and deeply penetrate a given social context, they become incubators for new cultural principles and arenas for new social games. The press, telephones, cars, the cinema, television – all of them passed through this stage. Now the same thing is happening for the computer.

Artifacts are ingredients in the development of a culture and, in industrial and post-industrial societies, technologists operate increasingly more openly as 'social activists' promoting an often Utopian but not necessarily less resolute vision of society (Kling, 1994; Dunlop and Kling, 1991). In the light of these reflections, attention paid to artifacts clearly does not allow us to stray very far from the perspective of situated action which we adopted at the beginning of this work. On the contrary, such attention emphasizes still further the symbolic dimension of the social environment, which is incorporated in both technological and normative artifacts.

5.2 ARTIFACTS AS PROJECTS INCORPORATED IN TOOLS AND NORMS

Artifacts do not modify just practices but also abilities, professional competences, and the ways of thinking which accompanied previous operations. They alter the very fabric of the environments in which preceding tasks were performed. When an organization introduces new technologies, it invariably finds itself having to cope with changes in its internal communications, its image and its structure (Fulk and Steinfield, 1990; Goodman and Sproull, 1990; Stinchombe, 1990). Even circumscribed technological changes can give rise to profound cultural changes; otherwise, we would be unable to see anything particularly dramatic in the coming of the Industrial Revolution and its sequelae which so transformed modern Western societies, and we would be unaware of the real nature of the changes which information and communications technologies are now introducing into our lives.

In her penetrating study of the changes induced by new technologies in the ways of working and living of blue-collar workers employed in some huge paper factories in the US, Zuboff (1988) notes that: 'The material alterations in their means of production were manifested in transformations at intimate levels of experience – assumptions about knowledge and power, their beliefs about work and the meaning they derived from it, the content and rhythm of their social exchanges, and the ordinary mental and physical disciplines to which they accommodated in their daily lives. I saw that a world of sensibilities and expectations was being irretrievably displaced by a new world, one I did not yet understand' (p.xiii). Tools not only modify the behaviour of actors while they perform their tasks, but also the social and physical environment in which the tasks were previously performed; they may even transform people's mentalities.

This will be seen to be true if we remember how the metaphor of knowledge as information processing, created with the development and spread of computers, has so profoundly influenced the ideas which human beings have about their minds. Contemporary cultural psychology (Cole, 1990) has strongly stressed the need to consider 'that people "think through artifacts" and that the shape of this thinking is constrained by the way the particular set of artifacts is put together as part of a historical stream of human activity, the motives for which are survival and reproduction' (p.287). Understanding of the true impact of technologies on human societies leads us not only to a more realistic view of artifacts as tools invented to perform particular tasks, but also to consider them as resources that allow us to transform our current cultural frameworks. Tools are not simply means for achieving local goals, they are also occasions on which to define new goals and find new meanings for both situations and actions.

The ideas that a given culture has, for example, of work, study or justice, are also artifacts, instruments used for interacting with the environment. They are embodied in social realities called workplaces, schools or law courts. And none of these is less solid or well-defined than, for example, knives, milling tools or lathes. In this wider sense, artifacts include not only the instruments required for certain physical operations on the environment, but also the social norms which establish what is to be done and how. The latter do not languish idly in the world of pure ideas, but are sometimes all too evident in the social and physical environment. They are manifest, for example, in the organizational structure of this particular company, the construction of that new wing to the school, or the management of that prison.

A normative social artifact holds within itself both the socially meaningful structure of an environment and its criteria for interpretation. For example, Taylor's theory of scientific management (1911) is at the same time the structure given to the management of the Ford workshops in the early decades of this century, and the organizational theory which interpreted the requirements (at that time) for mass production of cars so persuasively that even today many see the early Ford workshops as the incarnation of Taylor's principles. We noted in previous chapters that situations are always interpreted. We now stress that it is cultural order which forms the repertory of meanings available to actors, allowing them to interpret situations. The existence of this order provides a set of possible interpretations which are at leastly partially shared among actors.

Instrument-artifacts used to perform tasks and concept-artifacts formulating the basic principles of a given culture are both projects which have taken on a physical body, like a lathe, or a social body, like Taylor's theory of company organization. Speaking of artifacts does not therefore mean moving in an abstract dimension, far removed from actors' daily lives; it means placing on them the responsibility for technologies and social norms. Technologies and norms cease to appear as realities external to the actions of people to become what they really are, projects in which higher-rank artifacts dictate the way in which lower-rank ones are used. A large-scale organizational theory like Taylor's, for instance, establishes where a particular tool like a lathe must be located, how it must be constructed, and how it must be used.

In tool-artifacts, a tool cannot be separated from the task for which it was designed, and both depend on the project they serve. Those who design artifacts also design the ways in which the people who are to use them will act and interact (Winograd and Flores, 1986). A hunter's skinning knife, a fish-slice, or a butcher's cleaver are each designed for a specific use, and this is inherent in their very shape. A milling-cutter or a computer program are embodied projects in which the structure of the tool and the interpretation of possible situations in which it can be used merge.

Concept-artifacts are also physically and socially embodied projects, although when we commonly speak of artifacts, we usually refer to technological rather than normative artifacts. However, the distinction between tool-artifacts and concept-artifacts is neither clear nor solidly based. We have seen that concepts – for example, the way in which a car industry must be organized – may also be instruments by means of which actors model the social and physical environment. Ideas about how to organize a company or how to eat meat properly are no less practical or efficacious than the tools they shape, e.g. lathes or cutlery. In fact, in all cultures tools and norms merge, for the simple reason that tools serve to perform tasks, which in turn are performed in order to carry out more ample projects which are selected according to values, principles and norms which, again in turn, are formulated starting from the practical availability of proper tools. Although the operations of certain kinds of tools (lathes, cutters, etc.) may be controlled by a Tayloristic theory of work organization, such an organization cannot develop without those tools.

5.3   THE POLITICS OF ARTIFACTS: TECHNOLOGIES AND SOCIAL CONTROL

Artifacts are essentially the interfaces between actors and the environment. The overlap between them, which in Figure 1 (Chapter 1) defined the space for interpretation of both interests and opportunities, may now be delineated in more detail. The overlapping area may be envisaged either as the space of cultural order, which is the prerequisite for making shared meanings possible, or as the space where artifacts dwell, since artifacts embody the principles and values of a given culture. The two definitions are complementary: artifacts embody the principles of a culture no less faithfully than the engraved tablets of its laws.

Culture is a *corpus* of symbols, in its virtual and traditional functioning, and at the same time a *corpus* of artifacts, in its real everyday functioning. It would be incorrect to think that there is no overlap between meanings, which are supposedly located in actors' heads, and the social and physical environment, which is interwoven with artifacts. 'It is misleading to think that cultural conceptions must be located either outside the person or inside the person. In an authentic culture, cultural conceptions are likely assembled or reproduced in both places at once, and probably for good psychological reasons' (Shweder and Sullivan, 1993, p.511). Cultural psychology has kept this perspective alive even at moments in the history of social sciences when the divide between objects and ideas, between 'outside' or 'inside' the person, gaped at its widest.

Cole (1990) proposes a 'conception of culture as the unique medium of human existence, a medium that acts as both constraint and tool of human action' (p.282). Artifacts reveal the width and depth of this mediation (Cole, 1995). Thanks to artifacts, we can see that not only they but also other objects – not constructed, but 'found', as it were, in the environment – are the culturally negotiated responses of the environment to the interests and dispositions of actors. When Winner (1980) asked: 'Do artifacts have politics?', he was not simply trying to create an effect but pointing out that artifacts do deal with interests which are in many cases conflictual. The physical and social environment is an arena in which actors with competing interests and discrepant symbolic systems confront each other.

Research on the new technologies, in particular on the current developments of *human–computer interaction*, shows a growing awareness of this dimension of artifacts (Brown and Duguid, 1994). On the interpersonal–intergroup level, the political nature of artifacts is particularly evident in the oscillation between coop-eration and conflict in the development of the relationship between systems designers and users. The two groups have different social and professional inter-ests: the former try to produce instruments that will expand their professional skills; the latter expect that new technological products will fundamentally simplify their work routine and aid them in their usual daily activities (Mantovani, 1994a). Space for cooperation exists between the two groups to the extent to which each is interested in the development of the emerging technologies. But a conflict (which may be manifest at interpersonal and intergroup levels) also exists on such matters as control and power: while users can control the environment through the computer, designers can obtain the same results by controlling, even partially, users' performance (Tainsh, 1988).

So it is not surprising that cooperation between designers and users is often hindered by strategic games between actors playing different socio-professional and organizational roles. Bannon (1991) eloquently describes how, when starting work as a systems designer for a large organization, in a group working on human factors, he was asked to develop a new interface for an already operating system. With the aim of getting to the core of the problem, he asked to talk to some of the people who used the system daily. This was denied him. He writes: 'To my astonishment, I was informed that not only could I not proceed in this fashion, but – ultimate irony – I was not allowed, for political organizational reasons, to meet with even a single user throughout this process ... I complained, but being in a very junior position, I was overruled, so I then spent some months on what seemed to me an insane task: developing a logically coherent, though naive, mapping of tasks onto a menu-based interface based solely on the paper speci-fications I had been given' (p.26). This episode highlights the far from casual link between abnormal concern about ensuring 'logically coherent' formalization and the suppression of opportunities which designers could exploit to improve their knowledge of users' *real* needs.

In the perspective of situated action, as we saw in Chapter 2, the idea that actors' conduct may be totally controlled by prefixed plans is illusory, but it is a useful if not indispensable ploy for those who aim at social control through arti-facts. When considering the politics of artifacts therefore, we must bear in mind the 'political' dimension of formalism (see Chapter 6). Recourse to formal rules to affirm and maintain particular forms of social control is an ancient practice and one which is amply documented in the history of artifacts (Suchman, 1993). When rules are incorporated and hidden during the very design of artifacts, the exercise of social control is automatic, effective, but also indirect and difficult to identify and combat.

On the cultural level, the political nature of artifacts consists in grafting particular principles and values on to the structure of those artifacts, which thus act as only apparently innocent, neutral vehicles. In reality, there is nothing neutral about the 'rationality' and effectiveness of artifacts in technological ideologies (Ciborra and Lanzara, 1990; Dunlop and Kling, 1991; Lea, 1991). So artifacts are political objects in two senses. The first, more obvious but circumscribed, envisages them both as the battlefield and as the stakes to be won in the power games played by competing socio-professional groups. The second sense, deeper and more pervasive, is that 'artifacts take their significance from the social world ... at the same time that they mediate our interactions with that world' (Suchman and Trigg, 1991, p.74). As we saw earlier, artifacts are social projects, endowed with their own peculiar, definite meanings. The available meanings in the social context are not neutral but express the principles, values and choices which, in the end, involve actors' more or less explicit goals and identities. It is this second aspect of the politics of artifacts to which we now turn.

5.4   HUMAN RESPONSIBILITY AND THE SOCIALIZATION OF TECHNOLOGIES

The two essential questions of the politics of artifacts may be summed up as follows: who controls the products of a certain technology? and, what social project is incorporated in that technology? These questions have specific connotations as technological innovation steadily develops. In our opinion, they deal with human control over systems and the 'socialization' of new technologies (Norman, 1992). Let us address them in that order.

The current scenario, which includes the capillary spread of new information and communications technologies, presents a singular contrast. On the one hand, we have increasingly specialized designers, who live in a relatively isolated social and professional environment (Fishhoff, 1986) and who have stopped being representatives of their final users, although they may not realize how far they are now removed from the people who will eventually use the systems they design (Hammond et al., 1987). On the other hand, we have a great number of users who have no expertise in the new technologies and who generally have no desire to gain any. There is thus a paradox between technologies of increasing complexity destined for users with a decreasing level of preparation for using them (Suchman, 1987). Efforts to improve cognitive and social compatibility between actors and computer artifacts have concentrated on greater 'usability', and have been moderately successful, thanks to relatively efficient devices like icons, window menus and graphic interfaces.

But the basic problem remains. In new technological environments, actors often find themselves unable to interpret the information with which their artifacts supply them, nor do they have a clear idea of their responsibilities in the interaction with the artifact. One area in which problems related to responsibility and control are openly dealt with is that of the expert systems, which in many cases are presented as systems for aiding decision-making, tutorial systems, and the like. In their capacity as university teachers of expert systems designers, Speller and Brandon (1986) are in contact with the peculiar mentality of professionals working in this field. These authors call attention to the risks inherent in indiscriminate recourse to automatic systems as if they were capable of independent decision-making. 'New systems present dangers on two distinct levels. The first, and more direct, is where the system makes decisions and controls a process without human consultation. The second, and more subtle, is where humans religiously obey the system, implementing its decisions without understanding the significance and credibility of their actions. Either way, there ought to be a point where "expert systems" are accepted as assistants and not as oracles' (p.143).

New technologies tend to exclude actors' discretionary skills in decision-making processes, at least in certain situations (Nissenbaum, 1994). The resulting sense of loss of control over the environment is expressed by many who have experienced the sudden inrush of informatics technologies in their working lives. Zuboff (1988) quotes the words of an employee in a large American paper factory which had recently been fully automated: 'Doing my job through the computer, it feels different. It is like you are riding a big, powerful horse, but someone is sitting behind you on the saddle holding the reins, and you just have to be on that ride and hold on. You see what is coming, but you can't do anything to control it' (p.64).

To cope successfully with the problem of control over technologies, actors have to develop a cultural framework that is capable of explaining them properly, not only of using them in an acceptably effective way. Actors need high-level models of artifacts to help them interpret not just single segments of their interaction with them, but the entire new environment created by them. These environments are not easy to interpret, because the affordances to which our species has become adapted over hundreds of thousands of years of experience cannot be exploited in them. Countless generations have enabled us to respond promptly (and generally quite accurately) to the affordances offered, say, by a ripe apricot, a girl's smile, or a seat in the shade on a hot day.

This facilitation, born of adaptation, no longer helps actors in the artificial circumstances they now have to face. So we should not be too surprised if they feel threatened and confused by being, as it were, suddenly thrown in the deep end of environments whose overall meaning escapes them. One salient characteristic of the new technologies is that they may be incomprehensible at first sight, not simply because computers are somehow opaque (Brown, 1986; Ackermann and Tauber, 1990; Tauber and Ackermann, 1991) – a drawback users try to reduce by producing various types of mental models (Carroll *et al.*, 1988) – but also because the very nature of these technologies is intrinsically equivocal.

'Technology as equivoque' is the title of a study by Weick (1990), who teaches organizational behaviour at the University of Michigan. This 'equivoque' is something that allows various equally plausible interpretations and which is difficult to penetrate, being subject to misunderstandings and uncertainties of several kinds. For Weick, the new technologies are equivocal in the sense that '... they are simultaneously the source of stochastic events, continuous events, and abstract events. Complex systems composed of these three classes of events make both limited sense and many different kinds of sense. They make limited sense because so little is visible and so much is transient, and they make many different kinds of sense because the dense interactions that occur within them can be modeled in so many different ways' (p.2).

We see that the very complexity of new technologies has to do with interpretation of the environment, as situated action states. The new artifacts are both high-level instruments – 'cognitive artifacts' (Norman, 1991) which interact deeply with the human mind – and a source of considerable confusion, if they cannot be controlled at the highest level, which is that of the sense they confer upon situations and which they receive from them (Norman, 1993, 1994). Making sense of the environments of the new technologies is an innovative process, a discovery, which requires significant changes in the culture of both designers and users.

As far as systems designers are concerned, technical competence regarding artifacts must be grafted onto a more general comprehension of social processes, such as that supplied by the social sciences. At the present time, say Dunlop and Kling (1991), 'the dominant paradigms in academic computer science do not help technical professionals comprehend the social complexities of computerization, since they focus on computability rather than usability' (p.9). At the current phase of development of computer science, they believe, precious contributions may be made by the social sciences.

Conversely, users need to understand not only how the artifacts produced by the new technologies work, so that they can apply them effectively to the needs of specific ongoing situations, but also which social meanings are embodied in them and transmitted by them, in more or less transparent forms. In order to do this, actors must develop a new sensitivity towards the cultural and normative dimensions of artifacts. They must understand that the technologies are to be interpreted – and in fact already are, although surreptitiously, through the technological ideologies implicit in the current production and promotion of artifacts. Only then will they be able to evaluate critically '... the seductive equation: technological progress = social progress' (*ibid.*, p.5), which is still the credo of many communities of specialists working on the development of new tools.

It is within the framework of a new bond between situations and action, between symbolic order and effective practices, between social context and artifacts, that real collaboration between the designers and users of new technologies becomes possible. Stressing usability issues in order to arrive at effective cooperation is not enough. A shared interpretation of daily situations must also be developed, and that requires a common culture. Cooperation between various disciplines or socio-professional roles often breaks down precisely because there is no common ground for agreement. Successful *cooperative* design of tools, in which both designers and users work together (Bødker *et al.*, 1991), does not depend only on people's goodwill but also on the availability of sufficient *common* interpretative resources within the environment designed for that cooperation.

At the present time, actors do not seem to be fully aware of the extent of the changes that new technologies have produced, and will continue to produce in the foreseeable future, on social and individual experience. Many seem to think that all these new tools only supply means with which to achieve their previous goals. But situated action showed us that means and goals may become intertwined. We have also seen that changes in artifacts are in themselves changes in the symbolic system on which a certain society bases its values. For example, communication environments do not only represent an important background for the interpretation of the messages which pass through them, but are themselves meta-messages, as we shall see when we come to analyze virtual reality as a communication environment. Even less sophisticated artifacts like e-mail, although lacking the glittering fascination of virtual reality, oblige us to revise our ideas on communication as the simple transfer of information from one person to another.

PART THREE

New Environments of Cooperation and Communication

CHAPTER SIX

Computer-supported Cooperative Work

The cooperative ensemble is divided into myriads of small worlds with their own particular views of the world. (Schmidt, 1991)

Consider a round, sloped, multi-goal field on which individuals play soccer. Many different people, but not everyone, can join the game or leave it at different times. Some people can throw balls into the game or remove them. While they are in the game, individuals try to kick whatever ball comes near them in the direction of goals they like and away from goals they wish to avoid. (March, 1991)

6.1 DIFFERENT PERSPECTIVES: HOW CAN COOPERATION TAKE PLACE?

When we try to apply the situated action approach to environments braided with artifacts, we realize that we have to deal with cooperation and communication not in a general sense, but in real living and working contexts, characterized by peculiar relations among actors and also by those social productions inherent in them, which are called artifacts. However, when we seriously consider the presence of artifacts in everyday life, cooperation begins to appear more puzzling than we had initially thought. This is valid both for artifacts considered as tools constructed by new information technologies, which is how we will treat them here, and for artifacts understood as social principles and norms, whose effects on cooperation will be discussed in Section 6.2.

The contribution of artifacts produced by information technologies to the development of cooperative environments is currently at the centre of interdisciplinary studies, both technical and social, in a theoretical and applicational area called *computer-supported cooperative work* (CSCW). CSCW, an offshoot of research on human–computer interaction (HCI), is amply covered in journals such as *Behaviour and Information Technology*, *International Journal of Man–Computer Studies* and *Computer-Supported Cooperative Work*, in annual scientific conferences, and in publications discussing new theoretical approaches and the most interesting applications in the field. The works edited by Bagnara *et al.* (1994), Bowers and Benford (1991), Galegher *et al.* (1990), Greenberg (1991), Rasmussen *et al.* (1991), Robinson (1991), Stamper *et al.* (1991) and Wilson (1991) offer a stimulating picture of current CSCW prospects.

To CSCW goes the merit of having swiftly realized that trying to capture the reality of cooperation in real life by means of formal models is not an easy task. CSCW has helped us to become more aware of the social nature of cooperation, as HCI research helped us to understand more of the adaptive value of everyday reasoning. This is not surprising, because attempting to model cognitive and social processes is a blade that cuts both ways: on the one hand, through simulation, we can effectively test models of the mind and of social behaviour; and on the other, we are led to understand the richness of the processes which our models attempt to capture. A comparison between a given scientific theory and the context in which it is applied clarifies both the limitations of the theory and the relevant features of the context (McGuire, 1983). In this sense, scientific theories are also cultural artifacts – tools with the specific function of exploring the nature and characteristics of various contexts.

Artificial intelligence undertook a similar task when, required to reproduce and if possible improve the strategies of expert human decision-makers, it was faced with the difficulty of emulating some aspects of the peculiar flexibility of human capacity for diagnostics and decision-making. In so doing, it highlighted the then unsuspected subtlety of everyday reasoning and enlivened the debate on the features of mental models involved in it (Johnson-Laird, 1983, 1988; Bara, 1995). In the same way, CSCW studies are now discovering how sophisticated the mechanisms of human cooperation can be. In his opening speech at the second European conference on CSCW in Amsterdam, Schmidt (1991) stated: 'The innocence and familiarity of cooperative work is deceptive. Cooperative work is difficult to bridle and coerce into a dependable model. And anyone trying to incorporate a model of a social world in a computer system as an infrastructure for that world is as reckless as a daredevil mounting a Bengal tiger' (p.1).

Why is it so difficult to model cooperation properly? One reason is the ambiguity and unpredictability of situations. If we consider the world of daily experience as an open system – and it is difficult to deny that it is one – we must admit that 'it is impossible both in practice and in theory to anticipate and foresee every contingency that could occur during the development of a set of tasks. Therefore, no formal description of a system can be complete' (Gerson and Star, 1986, p.266). But it is a step we have to take. Gasser too (1991) says that, if we really want to think of multi-agent systems, we must develop theories that dissolve the distinction between open and closed systems, going as far as to 'consider all systems as fundamentally opened ones' (p.132). In terms of our context model, the justification of principle and not only of fact of the multiplicity of actors' viewpoints lies precisely in the inexhaustible richness of situations.

Any interpretation of a certain situation, however sensible it may appear within a given social and cultural context, is neither the only possible one nor necessarily the best: maps are not territories, even when they are good maps. Situations may always reveal unexpected hidden resources which are open only to those who can discover them using their imagination. Truly distributed decision-making environments (Rasmussen *et al.*, 1991) are such precisely because none of the actors in them is presumed to have either an accurate global vision of the problem or even the capacity and competences to decide alone on each aspect of the environmental situations to be controlled. It is the complexity of situations which requires diagnosis and decision-making to be distributed and negotiated among actors.

A second reason for the difficulty in capturing cooperation by means of a formal model is that actors may have utterly different perspectives. 'Two agents in principle cannot have *identical* representations', states Gasser (1991, p.119), '... simply by virtue of differing commitment histories and local circumstances' (*ibid.*, p.117). 'Shared knowledge is impossible', if by this expression we intend to designate 'several agents knowing the *same fact* interpreted the *same way*'. In Gasser's view, cooperation basically is not correspondence of coinciding representations but practical agreement, founded on the fact that 'we have ways of pragmatically aligning our activities and acting *as though* we share knowledge'. However, this appears to be an incomplete conclusion: pragmatics is only one of the two pathways which actors constantly follow, not the only one. Moreover, unless a repertory of shared meanings is not at least potentially available, how can actors agree on the sense to be attributed to their practices?

As we have shown in the preceding two chapters, actors do not have only pragmatic resources available to them in order to understand one another. There are at least two pathways, both used alternatively or even both at the same time. The first is *top-down*, from pre-existing cultural principles to practices; the second is *bottom-up*, from practices to the formulation of new cultural principles. Actors manage to cooperate to the extent to which, joining and mingling interpretations with practices, they are able to make plausible inferences regarding the meanings which they themselves and their interlocutors assign to situations from one moment to the next and which are changed by their intervention. At the same time, actors tentatively define in which world of principles both they and their interlocutors are moving. Beach (1990) rightly stresses that 'much of the social interaction that precedes group decision making is devoted to ironing out differences in the participants' frames through the sharing of information and through negotiation' (p.64). The core of social processes like cooperation and communication, i.e. the element allowing actors to develop shared meanings for actions, lies more in reciprocal recognition or agreement regarding the types of principles to be followed (the first level of our model) and interpretations to be given to situations (second level), rather than in the mere availability of information regarding the situation in question.

There are also culturally specific reasons which discourage us from presuming that different actors can spontaneously develop coinciding systems of representations in everyday life. We are now experiencing a peculiar moment of development in advanced post-modern societies in which the increasing cultural differences among actors – differences created by belonging to more and more specialized and self-sufficient professional communities, equipped with their own

values, knowledge, competences, interests, lifestyles and visions of the world –
are all manifest and enhanced. Even in academic circles, the cultural worlds of,
for example, physicists and doctors, are light years distant from each other. In fact,
even these categorizations are probably too ample and would be refuted by those
physicists and doctors as inappropriate. They would suggest that we should distinguish, for example, between astrophysicists and other physicists, or between
neuropsychologists and other kinds of psychologists.

Modern professions constantly tend to set themselves up as environments of
primary and sometimes exclusive socialization, in much the same way as territo-
rial communities, religious faiths and social classes in industrialized countries did
in the last century. In advanced, often multinational, work environments, individuals and groups 'use different vocabularies, have different motives, represent
organizations of widely differing cultures, and the referents of the transactions
may vary from highly abstract concepts to concrete products' (Williams and
Gibson, 1990, p.10). It may often be difficult not only to try to reconcile the various
points of view, but even to know what they consist of.

6.2   THERE IS NO OMNISCIENT CENTRE IN ORGANIZATIONS

It is not easy to understand how cooperation can originate and develop in the presence of complex, ambiguous situations on the one hand, and of actors hemmed in
by their own idiosyncratic perspectives on the other. Our model of social context
may offer a framework suitable to this problem, which is basic in studies on CSCW.

One explanation, traditional but unsatisfactory, of the possibility of cooperation was supplied by the structural theories on organization predominant in the
first half of this century, in which organizations were seen as structures divided
into ordered, hierarchical parts, reflecting the lines of command and control
(Fayol, 1949). These structures were considered able to guarantee the possibility
of cooperation (if coerced collaboration may be called such) since they were
believed to be capable of controlling and suppressing, at least temporarily, the
constitutional divergence of actors' interests.

The best-known of these theories was Taylor's (1911) *Scientific Management*,
which set itself up as the best or even the only rational approach to organization
and work. In reality, it only codified the constraints of a particular type of production on industrial relations. The growing car industry, by means of the prescrip-
tion of operations established by 'time-and-motion' offices, and through the
introduction of piecework and assembly lines, was obliged to exert strict control
on mass production in order to achieve coordinated mass production through the
assembly of manufacts composed of a vast number of standardized components.

Hierarchical models were based on the hypothesis that organizations contained
within themselves an omniscient and omnipotent centre (see Crozier and Friedberg, 1977, for a harsh criticism of this position). This centre was supposed to be
constantly aware of everything that happened in the organization and its
surrounding environment, and to be capable of controlling the movements of all

the actors inside it. According to this hierarchical model, it did not have the task of coordinating interactions among actors, but that of prescribing and controlling their correct performance. In Taylor's organizational conception, control and surveillance over employees were the basic functions of managers. Their main goal was not the reconciliation of actors' interests but their suppression, in favour of the viewpoint of the central authority.

The authoritarian character of these concepts should not mask their obvious utopian aspects. Hierarchical models presumed that conflict within the company would disappear if the centre was determined and able to carry out its proper task. They also fondly imagined that the environment would cease to be ambiguous and unpredictable if management had reliable sources of information and was sufficiently capable of processing it properly. These assumptions are now considered hardly acceptable in CSCW research. Schmidt (1991) assumes that cooperation is mainly based on discretionality and negotiation among actors, which is mandatory if we accept that 'the cooperative ensemble is divided into myriads of small worlds with their own particular views of the world' (p.6). CSCW studies refute the latest appearance of Taylorism in the form of *office automation* as futile and unrealizable.

Organizations are now seen rather as open systems (Jirotka *et al.*, 1992) trying to cope with continually changing environments by means of various forms of negotiation of interests among the various social actors composing them (for negotiation, see the review of Carnevale and Pruitt, 1992). They are considered as *networks*, as suggested by the approach of the Tavistock Institute to socio-technical systems (Emery and Trist, 1969; Ehn, 1988), as *organisms responding adaptively* to unstable environments, according to contingency theory (Williamson, 1986), or as *units of information processing* and decision-making, following the traditional decision-making approach (Simon, 1960), although this has recently come under heavy fire from March (1991). Organizations can also be conceived of as *cultural constructions* in an ethnographic sense (Gilbert and Mulkay, 1984; Latour, 1987), centred on identity problems, careers understood as rites of passage, and so on. Or, finally, they can be seen as *productions* in an ethnomethodological sense (Lynch, 1985), analyzed starting from the tiniest daily practices, details of conversations, gestures, etc.

These various organizational theories are all in competition, influencing the life of organizations and our understanding of cooperation. A theory of organization is an artifact, not a 'natural' thing and, like all artifacts, it has a goal. The goal of the old theories of organization was social control over employees' work. More generally, actors were to be reassured that conflict could not prevail and that superior rationality, mystically incarnate in management, understood and could maintain social and cultural order.

These days we are less worried about maintaining this order, or rather – what is more plausible – our ideas about how an acceptable social order should function are more elastic than those of our predecessors. As a consequence, we are now prepared to tolerate this quite drastic statement by Schmidt (1994): '*The* organization does not exist. The organizational phenomenon is a complex of superimposed and interacting organizational formations ... Further, there is no overarching conceptual scheme for the analysis of organization – and none is required. No single generative mechanism can explain the formation of organizations in general nor the formation of firms' (pp.109–110).

Modern Western cultures attribute particular symbolic value to the 'rationality' of decision-making processes and to the related need for highly accurate information management. For this reason, their view of organizations aims to 'reassure those involved (a) that the choice has been made intelligently; (b) that it reflects planning, thinking, analysis, and the systematic use of information; (c) that the choice is sensitive to the concerns of relevant people; and (d) that the right people are involved' (March, 1991, pp.110–111). In reality, things are not quite so simple. March gives us another image, both witty and realistic, of how decisions in organizations really *happen* in everyday life. 'Consider a round, sloped, multi-goal field on which individuals play soccer. Many different people, but not everyone, can join the game or leave it at different times. Some people can throw balls into the game or remove them. While they are in the game, individuals try to kick whatever ball comes near them in the direction of goals they like and away from goals they wish to avoid' (*ibid.*, p.108). What does *collaboration* mean, in such a multiple and highly creative soccer game? There is no central coordination with which the various players can cooperate. There are not even two teams playing. So it is impossible to recognize any kind of overall plan or durable scheme during the game. Nor can we say that there are as many teams as there are players, because more or less stable and inclusive or exclusive coalitions form continually, according to the balls, goalposts and reciprocal positions of the players.

If, at one moment in the game, actor X exploits a chance offered to her or him of kicking the ball towards a certain goal which is also of interest to actor A, thus disturbing actor B and damaging actor C, with whom she or he happens to have certain aims in common (although at this particular moment they are in abeyance), can we say that X is cooperating? With whom? How can we distinguish cooperation from conflict and from negotiation, in contexts with multiple decision-makers and ever-changing goals like those of everyday life? These ideas are completely extraneous to structural concepts, which see organizations as compact units devoted to the dispassionate processing of information and the formulation of carefully worked out decisions.

However, we should not believe that the concept of an organization as an arena in which various interests conflict inevitably leads to a state of chaos in which social actors blindly follow the impulse of the moment, incapable of stable, participatory plans. There *is* an order, but it is neither rigid nor necessarily descended from the centre: 'the ways in which organizations bring order to disorder is less hierarchical and with less means–ends chains than is anticipated by conventional theories. There is order, but it is not conventional order' (March, 1991, p.109). Our idea is that order is not basically inherent in hierarchy, but in the development of a shared meaning, of which the hierarchy may be an agent. Symbolic interactionism maintains this position: 'The basis for interaction among organization members is a shared system of meaning. People have developed many shared assumptions and understandings about the meaning of words, actions and events' (Trevino *et al.*, 1990, p.73).

If shared meaning does not exist, or does not yet exist, the situation is confused. This is not only disturbing for the actors involved; it also gives rise to continuous confrontations. Conflicts may be lessened or avoided if appropriate responses are negotiated on the symbolic plane: 'In ill-defined situations, people must create a common understanding before they can make decisions that others will comprehend, agree upon, and accept. In this type of situation, organization members proactively shape reality together. Through negotiation and feedback they decrease ambiguity and create symbols that establish new organizational meanings' (*ibid*.). When the symbolic framework is finally in place and works reliably, cooperation flows smoothly: members of organizations do not need to negotiate the meaning of actions and events every time, if they can count on reciprocal understanding and on a certain amount of consensus right from the beginning.

We do not have confused, ill-constructed situations on the one hand, and clear, well-structured situations on the other, as we could expect according to rationalistic decision-making theories, which blame the ambiguity of everyday environments on the disorder existing inside decision-makers' heads. We have to deal with more or less shared, appropriate symbolic systems which activate equally shared, appropriate interpretations, by means of which actors can experiment with more or less clear-cut situations. The current disorientation regarding the social and cultural implications of technological innovation reveals not so much the presumed immaturity of new technologies and their still experimental character, as is often said, but rather the absence of appropriate cultural responses to questions posed by the current forms of the technology–organization mix.

It is on the construction of an appropriate cultural framework that possible improvements in cooperative work in technologically advanced environments in the near future depend; not merely on a process of spontaneous, gradual maturation inside technologies which, left to themselves, would eventually find their own appropriate ways of functioning. The two viewpoints lead to different consequences. In the first case, the entire social context is seen to be invested with innovation, and changes as a result of the introduction of new technologies. Thus, 'computer socialization' (Ellis, 1991; Kling, 1995) is not only an option of interest to social sciences researchers, but is also a challenge that postmodern cultures must overcome in order to maintain their control over living and working environments. In the second case, not only do we give a local, reductive meaning to the changes induced by new technologies, but the entire question is left in the hands of professional technicians who, if they are left alone, are not properly equipped to manage successfully the social and cultural issues emerging in the new environments of cooperation and communication.

6.3 THE POLITICS OF FORMALISM: DISCIPLINING ACTIVITY

The ambition of techniques to organize work environments is not a novelty but a salient characteristic of the modern age, in which 'industrialization created both the need and the possibility for purely *technological* control – as opposed to either organizational or individual human control – of physical movements and flows' (Beniger, 1990, p.34) of production and activity in work environments. In previous times, the definition of the symbolic order and social control necessary to ensure respect by actors were the prerogatives of political and religious authorities and, later, of particular politico-moral and professional organizations. During the Industrial Revolution, workplace control was substantially removed from the hands of traditional authority and managed as the exclusive competence of company directors and, on their behalf, by the owners. As soon as technologies became sufficiently reliable and fast in processing information, they appeared as the best way of affirming power and authority in the workplace. Technical personnel were ready to assume functions of control over lesser employees and to incorporate those functions into technological artifacts to be used in ordinary activities – in exchange for a seat on the board.

Modern technologies, especially in the sphere of informatics, are well equipped to undertake functions of social control. 'Because of the abstract and highly generalizable properties of control, technologies for the control of material flows can be readily adapted to the control of information and humans, and hence to control within organizations' (*ibid.*). However, the price to pay for managing actors through technologies is high. People have to be considered as things, mere technical objects: 'The reason why people can be governed more readily *qua* things is that the amount of information about them that needs to be processed is thereby greatly reduced and hence the degree of control – for any constant capacity to process information – is correspondingly increased' (*ibid.*, p.39). It is only on this condition that technological control can hope to function.

Technological control is often prepared to pay this price. It no longer sees autonomous, independent actors as people at work. It even tries to convince them to stop thinking of their own interests and goals, insinuating that it is not really a good thing to cultivate personal interests in potential or actual contrast with those of the company. For those who have a clear idea of the close connection existing between technology and social control in advanced work environments, the 'politics of artifacts' is not just a tag but a substantial piece of evidence that is readily discernible all around us.

One particularly effective and widespread form of incorporation in artifacts of control functions which compete for power and authority is what Bowers (1992) calls 'the politics of formalism'. Let us designate formalism as 'a representational system of a certain sort. A formalism generates representations through the operation of rules over some vocabulary. The elements which make up the vocabulary and the terms which constitute the rules may represent human or machine action, computational procedures or operations, etc.' (*ibid.*, p.234). This allows us to separate distinct, countable elements, to see complex operations as a composition of simple components, to give temporal order to actions to be performed and, finally, 'to distinguish between the legal and the illegal. Only some representations are allowable, only some orders or compositions can be realized or recognized in a particular formalism' (*ibid.*). So formalism defines an order, priorities, a principle of discrimination between what is legal and what is not. In this sense, the assumption in a cooperative system of a formal model coincides with the establishment of a control function.

Some cooperative systems which rely most heavily on formal models make no mystery of their interest in controlling users' activities. Flores *et al.* (1988), referring to *The Coordinator* cooperative system constructed by them on the basis of one interpretation of Searle's theory of linguistic acts, state: 'We are primarily designing for settings in which the basic parameters of authority, obligation and cooperation are stable' (p.173). They explain that '*The Coordinator* has been most successful in organizations in which the users are relatively confident about their own position and the power they have within it. This does not mean that the organization is democratic or that power relations are equal. It means that there is clarity about what is expected of people and what authority they have' (*ibid.*). The clarity and stability in work environments required by Flores and co-workers as a condition for their system is quite different from the ambiguity and changing state typical of everyday situations. Suchman (1993) comments: 'Rather than being a tool for the collaborative production of social action, *The Coordinator* on this account is a tool for the reproduction of an established social order' (p.10) Suchman's position as well as Winograd's (1994) response to it prompted a debate which appeared in the 1995 spring issue of the journal *Computer-Supported Cooperative Work*.

Although we now have cooperative systems which try to preserve space for negotiation between the actors involved in workflow (Medina-Mora *et al.*, 1992; Marshak, 1993), it seems that, in CSCW systems, it may prove difficult to separate functions favouring cooperation from others which make control instruments out of such systems. This fact is immediately apparent if we analyze, for example, Schneider and Wagner's (1993) description of an information-sharing system for doctors and nurses within a large hospital. The system, called *Dossier Représentatif*, explicitly undertakes tasks of social control. 'For nurses who use it in their everyday practice the system offers help in keeping track of their responsibility for all patients. At the same time it provides a framework within which to describe and legitimize nursing interventions' (p.247).

Legitimizing practices, prescribing interventions, and keeping records on operations undertaken are precisely the functions of power in organizations. The *Dossier Représentatif* is not a simple organizational tool, like service orders or periodic personnel meetings. It is much more. It is power and authority both operating and at the same time concealed within an interactive system: 'While the design of the system initiated an important process of defining norms and routines, its use establishes a certain degree of discipline and rigour' (*ibid.*). Discipline, rigour: here are the very words that are used to describe hierarchical concepts of organizations. But in this case, the hierarchy is disguised on the one hand as an aid, and on the other as technical rationality. After all, we may think, to keep proper documentation of our acts may be useful. But – and this is the crux – who really possesses the information? If this is not in the exclusive possession of the actor, it is a potential threat to his autonomy. Information on his activities may be used against him, unbeknown to him, or for purposes extraneous to him.

In order to modify power relations, to reduce actors' autonomy and to control their activity, information does not necessarily have to be *used*. The only two requirements are that it be available and ready for use, and that actors should know this. Who controls the information, improperly defined as 'shared', in the *Dossier Représentatif*? Who actually keeps it? Who can make strategic use of it, if they have the power and capability to do so? These problems should not be minimized or deliberately kept shadowy. Information in organizations, states March (1991), is rarely innocent. Information is the raw material of power and lies at the very centre of power games played by people and socio-professional groups.

It is also becoming increasingly clear to CSCW researchers that while formalism has some advantages, it also has a price. 'No-one would dispute the power of formalization. Equally, no-one would dispute that there is, or can be, a price to pay for formality' (Anderson *et al.*, 1993, p.1009). We must become more aware of exactly what kind of price some types of formalization require, and carefully evaluate possible alternatives. Bowers (1992) hopes for the development of 'formalisms [which] can be used as resources for action, rather than as specifications for action' (p.256), which do not reduce actors' discretional space, which are distributed, and so on. This perspective is undoubtedly stimulating, but it requires designers and social actors to develop sensitivity and culture which are still lacking, except in particular contexts. At the present time, formalisms of this type are difficult to find.

What we now see more often in work environments affected by new information technologies is a concurrence of formalism and bureaucracy: 'Formal organization and rationalization intersect in the regulation of interpersonal relationships in terms of a formal set of impersonal and objective criteria' (Beniger, 1990, p.38). With all our new technologies, can we not produce anything better than this? What must cognitive systems be like in order to support human cooperation without imprisoning it? Our answer is that we should recognize that cooperation is always situated, always conditioned by specific circumstances and by actors' special interests. So any model of cooperation embodied in a cooperative system must be considered as a reference framework whose validity is limited and intrinsically 'precarious' (Schmidt, 1991). Although we must have models in order to construct cooperative environments, even in the best cases, any model can only be valid inside a restricted area of application. Even inside this area, it only captures those aspects of the application which have been modelled congruently, with respect both to the characteristics of the situation and to actors' goals, which may change from one moment to the next.

In a cooperation model, what is critical is not so much the amplitude of possible applications as the capacity to make actors understand promptly when the interaction is no longer suitably guided by the model. This is particularly important in models used in cooperative systems, which by definition have many users. Users may not be able to realize, sometimes for quite long periods of time, that the system is not functioning properly. This is because feedback to single actors is always local and does not allow them to understand whether the responses of the system as a whole are suitable to the ongoing situation. Each actor only sees one little piece of the system's outcome, and cannot assess its overall performance or even less its congruence with respect to the peculiar situation to be managed.

We should also recall that each model can only be applied to the extent that it adapts to the type of cooperation which is effectively relevant at that particular moment for those particular actors. From this standpoint Mantovani and Bolzoni (1994), using the 'cognitive walkthrough technique', analyzed three multi-actor, multi-goal vocational guidance systems (VGS) developed in Northern Italy to back up counselling in a careers advisory service. Their study first examined the relations between the various types of interactivity allowed by each system and the various goals of the social actors involved. It then considered the interactive features of the counselling sessions offered by each of the three systems, relating them to the contribution furnished in each of the design phases by the actors – defined in terms of their social roles – involved in the intended counselling activity. They found that open negotiation taking place during the early phases of system design between different social actors involved in the use of VGS allowed clarification of the strategic choices in modelling these complex multi-agent, multi-goal environments.

Some quite satisfactory examples of cooperative systems do exist. Robinson (1991) praises the GROVE system (Ellis *et al.*, 1991a), constructed to stimulate cooperation among several authors co-writing a text, and states that it functions quite well because it can tolerate a certain degree of ambiguity while still maintaining sufficient clarity: 'In general it can be said that any non-trivial collective activity requires effective communication that allows both ambiguity and clarity' (p.43). Cooperation, like communication by means of natural language, needs both clarity and ambiguity because it must capture in a single framework the different points of view, interests and values expressed by the various actors participating in a given social context.

We say 'cooperation' for the sake of brevity, but it should not be forgotten that *cooperation*, *conflict* and *negotiation* are all inextricably intertwined. Cooperation denotes not an improbable state of grace in which actors interact in perfect accord, but a phase of that chaotic but creative soccer game which is interaction in daily experience. 'These ideas of ambiguity and clarity can be developed as the 'formal' and 'cultural' aspects of language as used by participants in projects and organizations. ... Applications and restrictions that support one level at the expense of the other tend to fail' (*ibid.*). This happens because, in cooperative systems, 'the formal level is meaningless without interpretation, and the cultural level is vacuous without being grounded' (*ibid.*) in a specific social context. The affinity between these positions and the ones we have developed is clear. CSCW now faces the challenge of producing cooperative, not coercive, systems for actors. If it can assimilate the cultural dimension of action, it may supply artifacts which are not only more usable in real-life contexts, but also more appropriate to them.

6.4   SUPPORT SYSTEMS IN GROUP DECISION-MAKING

It should first be noted that 'cooperative work' is not the same as 'group work', as Bannon and Schmidt (1991) rightly stressed when refuting the proposal of Greif (1988) to consider 'CSCW' and 'groupware' as equivalent. *Cooperation* is a wider concept than *group*. In its current acceptance, *group*, which denotes a set of people who interact, possibly face to face, is only one of the forms by means of which actors may develop their ploys of cooperation, conflict and negotiation. This statement, which will be taken up again in Chapter 7 on communication, plays its part in the economy of this discussion. Our aim is to emphasize that social contexts cannot be reduced to interpersonal relationships, which are a part of the social world, not the whole of it.

Groupware systems have been developed mostly without a precise conceptual model of the social processes that they were presumed to support (McGrath and Hollingshead, 1993, 1994), mainly because priority was given to the technical problems involved in design and implementation rather than to issues concerning the expected final use of the systems. Attention to the social side of groupware tools has recently been paid by Grudin (1994), Huber *et al.* (1993), Hollingshead *et al.* (1993), Kottermann *et al.* (1994) and Walther and Burgoon (1992).

A significant effort to offer an orderly categorization of this fuzzy field has been made by Ellis and Wainer (1994), who propose a goal-based model of collaboration as a framework for understanding working group activities. They also build a functionally based taxonomy of the main groupware components: 'keepers', who manage shared information spaces (e.g. an object-oriented database); 'synchronizers', who control the flow of group activities; 'communicators', who support human-to-human communication such as e-mail; and 'agents', who perform specialized activities within the group, like the 'kitchen critic', a software tool embedded in a larger building-design aid system for supporting teams of architects in their work. The study of Ellis and Wainer aims at enhancing groupware products by bridging the gap between the goal-directed activities of social actors and the designers' formal models of groupware systems.

Let us now consider what kinds of artifacts new technologies have developed to sustain cooperation. First come *shared spaces*, *common information spaces*, or *community handbooks* for special work environments. There is not much cooperation here. If we recognize that the viewpoints of various social actors are and must be unique, they clearly cannot be reconciled simply by creating storehouses of information to which everyone has access. Information is not knowledge; nor is it the interpretation of situations, the meaning attributed to single acts and events.

Information is the resource which actors use in their ploys, but it is neither the game nor its rules. Schmidt and Bannon (1992) state that 'it is important to realize that one cannot just produce a common information space, that it does not automatically appear as the result of developing a common dictionary of terms and objects, as the meaning of these terms and objects must still be determined locally and temporally' (p.28) by actors.

When two or more actors cooperate, what is really shared is not information but its meaning. Cooperation is not a shared space: 'a uniform, complete, consistent, up-to-date community handbook is simply a chimera' (Schmidt, 1991, p.12). In order to create a shared meaning, information must be interpreted by each actor, and the interpretation which each one of them gives must then be socially negotiated. It would be naive to assume that cooperation coincides with the availability of a relational database filled with supposedly neutral and freely accessible information. We know full well that information is never neutral, since it is labelled and interpreted according to its provenance, its presumed aims, and its internal organization.

Information is normally used by actors for strategic purposes, i.e. as a resource for attaining their own goals. For example, we have seen that a common information space like the *Dossier Représentatif* may function as a powerful instrument of social control. The socio-professional conflict is endemic in work environments. It is a constitutional fact, derived like so many others from the ineluctable differences of actors' interests and goals. It is an illusion to believe that the existence of common information spaces can annihilate socio-professional competition and automatically stimulate the development of favourable reciprocal attitudes among actors. The occurrences of 'electronic altruism' (Sproull and Kiesler, 1991) – consisting mainly in the fact that people who carry out similar tasks are sometimes willing to exchange practical suggestions about work questions – may be viewed more as the effects of intra- and intergroup processes and reciprocal aid inside homogeneous socio-professional groups, rather than as peculiar positive effects of information technologies.

The spread of common information spaces has in fact no magic effect; on the contrary, it may hinder actors' initiative and creativity. March (1991) notes this danger in the excessive adoption of new technologies in organized work environments: 'Strategies or technologies that improve the sharing of knowledge, information and experience (e.g. education, databases) are very likely to do more for exploitation than for exploration'. This happens, he says, because 'they often reduce experimentation and increase reliability. Any technology or procedure that improves the reliability of behavior greatly while its average intelligence only slightly, is likely to be disadvantageous in the highly competitive situation where relative positions matter' (p.113). The risk is that the ways in which actors think of and cope with problems may be reduced to unrelieved flatness. If this were one of the results of the adoption of cooperative systems, we could say that they give wrong responses to problems posed by the multiplicity of actors' expectations. The right response, as already stated, consists in negotiating the various perspectives within a shared, meaningful framework (McCarthy and Monk, 1994a,b; Whittaker *et al.* 1993).

Apart from common information spaces, systems for cooperation in the CSCW area are mainly destined to intervene in meetings, which are regarded as highly deficient social activities because they are so often chaotic, unproductive, poorly documented and in need of 'help' from the new technologies (Poole and DeSanctis, 1990). Going back to the 'funny soccer game', we may think that the apparent disorder which is so characteristic of meetings conceals a deeper social order and perhaps even efficient negotiation, and that the idea of introducing order through technologies is dangerously close to the plans for control and discipline underlying systems like the *Dossier Représentatif*. Nevertheless, the idea that technologies can function as a sort of crutch assisting the decision-making processes of actors who are not fully rational in their usual conduct transpires from the very name given to these systems – *group decision support systems* (GDSS).

GDSS apply technologies from three main areas to the aim of improving human decision-making in groups and organizations: they are tools for decision-making, communication networks and advanced computer interfaces (see McLeod, 1992, for a review of ten years of experimental studies on synchronous group support systems). Decision-making technologies supply tools for modelling decisions (e.g. decision trees), methods for structuring groups at work (like the Delphi techniques) and rules to organize group discussion (like the so-called parliamentary procedures). Communication technologies produce electronic systems to transmit messages; these are true jewels of efficiency, apart from their interfaces, which are sometimes rather poor. They also provide such things as teleconferencing and allow other forms of preserving and transmitting the collected information. Computer technologies offer multi-user systems, enormous graphic capacities, and very advanced interactive languages.

This *corpus* of artifacts is supposed to come into contact with group interactive and decision-making processes but not to control them too overtly. In theory, the groups themselves should decide how to use these technologies and how to fit them into their scheme of things. In this sense, every GDSS in use should be considered as the product of a 'social technology' which actually shapes its real function (Poole and DeSanctis, 1990, 1992; McGrath and Hollingshead, 1993). However, in another sense, the technological artifacts that make up the hardware and software of GDSS do influence group processes, if only by introducing precise rules and clear procedures; for example, voting on proposals made within the group. They also make available homogeneous, fixed resources, e.g. the public display screen that the group uses as a shared work space.

According to Poole and DeSanctis (1992), in GDSS 'the structures provided by the technology are fairly easy to discern' (p.9) because they are visible to the naked eye. Fewer inferences by actors with respect to everyday situations are required, and this makes it easier to identify the functioning of the systems and the structuring of the group. However, although GDSS allow actors to participate in social interaction in a more controlled and simplified field with respect to everyday reality, the above authors do admit that actors' control of systems may meet with restraints of various kinds, due to the particular nature of contexts, lack of time, or users' incompetence. Above all, in spite of the order which GDSS wish to introduce, 'social systems and their interconnections are often so complex that users cannot fully grasp the implications of their actions' (Poole and DeSanctis, 1990, p.181).

In reality, what often happens when new technologies are introduced into cooperative activity is not simplification but great complication, because the new technologies are far from being crystal clear to their users. Not only do they not immediately disclose how they work, but they do not reveal their intersections with the social organization of the current activity which is, in turn, complicated by the presence of artifacts that are not easy to control. Weick (1990) stresses that interactive complexity is an 'an indigenous feature of new technologies'. This arises from the fact that 'the artificial restriction of perception and action by directives does not stop elements of the technology from interacting, but it does curtail comprehension of these interactions, which means that the technology becomes more complex and more puzzling' (p.37).

Faced with the artifacts produced by new technologies, actors are obliged to try to give them some meaning: 'Material artifacts set sensemaking processes in motion; sensemaking is constrained by actions, which themselves are constrained by artifacts; and sensemaking attempts to diagnose symptoms emitted by the technology' (*ibid.*, p.33). The meaning of a given situation to an actor is not inherent in the artifact itself, but must be imagined by correctly interpreting the clues that it lets fall. Faced with a GDSS, actors have two problems to solve instead of one: the old problem of working within their group, and the new one of making sense of the situations mediated by the new technology and using it appropriately.

Clearly, there are as many solutions as there are real contexts for using a GDSS. This is why the literature on the results of cooperative systems shows such extraordinary discrepancies: the quality of decisions taken, the equality in participation among group members, and the degree of approval towards the new artifacts all vary according to the characteristics of any given situation and to the different capacities of groups to control artifacts. The main consensus in research regards the greater time costs required for negotiation in electronic environments.

Let us now leave the features and kinds of cooperative systems, whether they be common information spaces or GDSS, and pass on to consider the impact of the new artifacts on the internal organization of activities in working groups. It is plausible to suppose that, if information flows are changed, the new technologies will also alter the times and modes of social interaction. We are not referring to time in the sense of the myth of efficiency, according to which we should always tend to do more things in a shorter span of time, which inspires the system rationalism criticized by Lea (1991). Time here is understood as the rhythm of interaction (Warner, 1988; McGrath and Kelly, 1986), the ebb and flow of sounds and silences in conversation. This perspective was adopted by Hesse *et al.* (1987) to compare the different structures of temporal organization prompted by CMC and face-to-face communication.

McGrath (1990) asks whether the new technologies do or do not alter the temporal aspects of work and communication in groups: 'The answer', he says, 'is a loud and definite "yes". The effects appear to be both powerful and pervasive, and they contain both positive and pernicious consequences for group behavior' (p.52). For individual actors, as we know, real-life situations often involve problems of interpretative ambiguity, of conflicting interests in the management of time, of scanty attentional resources. For groups, which must reconcile the perspectives and activities of more than one actor, these problems are multiplied. The needs, implicit in all group activities, of coordinating many different actors augment both the ambiguity of the situations and the potential for conflict among group members.

Groups face temporal problems by programming the things they have to do and the meetings they have to hold, paying special attention to the synchronization of their actions and negotiation on the allocation of available resources. Group members respond to the same problems by taking on commitments in terms of timing, negotiating rules and the priority order of behavioural sequences, regulating the flow of activities and of interpersonal interactions. Each of these problems, both personal and group, is directly influenced by new technologies. Imagine, for example, how many instruments have been produced to manage agendas and meetings. It is actors, not technologies, who control flows. But the technologies do change the forms of interaction, allowing new ploys within group life. They change the production function of groups, not only in the routine execution of tasks, as in the already-mentioned task–artifact circle, but also in the supraordinate activity of plan generation.

The production of a group plan involves a sequence of stages which begins with identifying a goal, continues with choosing means and the course of action to be followed, and concludes with implementing the actions necessary to achieve the goal. Two phases in the sequence are indispensable for the plan to exist: the first and the last. The intermediate phase, although not imperative for the birth and achievement of the plan, is essential for its quality because it is in this phase that greatest attention is paid to the search for really satisfactory solutions to problems and conflicts. If there is not much time to generate and complete the plan, negotiation of conflicts among group members and consideration of alternative courses of action are sacrificed, wholly or partly, to urgency (McGrath, 1990). This can have strong negative effects not only on the production function of the group but also on the other two functions, group well-being and support offered to members, both of which depend on the favourable outcome of negotiation among actors.

The importance of negotiation for good group functioning is confirmed by the study of Kraut *et al.* (1990) on strategies adopted by scientists committed to advanced experimental research to extend their group by enrolling new members. In choosing a new colleague with whom to work (which is considered by those involved as an operation analogous to that of choosing a companion with whom to live) scientists spend much time attempting to establish efficient synchronization, which is possible in physical proximity and using high-quality communication channels; i.e. working accurately in both directions and exploiting more than one sensory channel at a time. Especially during the first steps of a collaboration, it seems to be essential for people to rely on a flexible system of communications allowing mutual fine tuning, both on the emotional plane and on that of the image of the self. From this viewpoint, electronic communication may not be particularly suitable as the main channel to be used in the creation of new groups or in the initial phases of the development of important and innovative plans.

CHAPTER SEVEN

Computer-mediated Communication

▌

The theory of communication as information transfer separates knowledge from communication; it treats knowledge as an object that exists independently of the participants in the innovation venture. With this independent existence, information becomes an object that can be carried through channels. (Dohény-Farina, 1991)

Rather than viewing the organizational consequences of Information Technology (IT) as being inherently positive or negative, IT is best viewed as neutral regarding its organizational consequences. The nature of an organization's IT use emerges through complex interactions among the intentions of key actors, attributes of the technology involved, and dynamic organizational processes. (Zmud, 1990)

7.1 BEYOND THE MODEL OF COMMUNICATION AS INFORMATION TRANSFER

Computer-mediated communication (CMC) is not a novelty: the first large computer network, ARPANET, was developed in the late 1960s. Since then, computer networks have developed to an extraordinary extent and at lightning speed. The novelty of CMC lies in its current spread, which has thrown such a blaze of light on the new environments created by electronic communications. Now that the new technologies have, as it were, flexed their muscles and revealed their capacity to modify traditional forms of communication (Schuler, 1994), the social sciences are increasingly interested in understanding the characteristics of CMC and its effects on people, groups and organizations.

Asynchronous CMC, like e-mail, allows its participants to go beyond the barriers of both space and time. It permits people to communicate with other people when they choose, leaving messages that receivers can read when they open their e-mail boxes. Here, participants need not be present contemporarily, while synchronous CMC, such as teleconferences – with or without voices, with or without speakers' images on the screen – requires interlocutors to be reciprocally present at the same moment during the interaction. Hence problems may arise, in particular as regards turn-taking. In face-to-face communications, these problems are easily solved by social norms and negotiation; in CMC they are more difficult (Egido, 1990; Weedman, 1991). Asynchronous CMC is widespread and relatively homogeneous in technologically advanced countries, substantially reflecting their degree of development and economic integration. Synchronous CMC has developed less rapidly than its promoters had hoped, particularly in Europe, where some restrictions have been imposed by dissimilar governmental regulations.

The differences between CMC systems make it advisable to consider the various forms of CMC separately. For example, in teleconferences, the particular type of system used may constrain the interaction in its style as well as in its content, so that the results of research on one system in one context may hardly be compared with those on another. E-mail does function quite uniformly, although diverse interfaces are used. However, differences in organizational cultures and professional figures do make themselves felt, so that different social contexts of CMC use – ranging from the physicists at CERN in Geneva to the employees of IBM or the Rand Corporation in the United States – cannot easily be considered as homogeneous.

At the present time, apart from the peculiar environments of artificial reality or virtual reality (see Chapter 8), one considerable constraint common to a large part of CMC is the fact that texts have to be typed on a computer keyboard. Exceptions are videoconferences which, although they have been available for about 30 years, were seldom used until recently (except by American Telephone & Telegraph; Egido, 1990). Video communication systems usually use voices and images, with some limitations. Particular applications have recently been developed, like the audiovisual network in the EuroPARC laboratories of Rank Xerox in Cambridge, in which a stable video link among various work environments is permanently open (Heath and Luff, 1993).

Some problems arise in CMC environments because of the fact that messages are in written form. Writing is costly in terms of time, because it takes far longer than speaking. Users try to avoid this difficulty by sending cryptic or condensed messages, but the quality of the communication may suffer. Pieces of information may be lost, sometimes all the connotative aspects of messages – inherent in ordinary letters and to an even greater extent in *viva voce* messages, which contain so much information about the writers and their intentions – may also disappear.

Although writing is prevalent in CMC, its social significance is quite distinct from that of non-electronic written communications, i.e. letters sent by ordinary or internal mail (Perse and Courtright, 1993; Rice, 1993). Lea (1991) conducted a study on the way in which users perceive the various means of communication, and found that CMC is perceived as a separate medium completely different from traditional methods. It therefore seems sensible to consider CMC as a communication environment in its own right, rather than as a particular case of written communication.

The meaning attributed by actors to the choice of the media most suitable for different types of communications is considered by Trevino *et al.* (1990) in a symbolic interactionist perspective. They observe: 'In the past the medium of communication has been conceptualized as a simple pipeline, a carrier of messages. ... This pipeline is selected for convenience, availability, or for its capacity to transmit a certain kind of message. However, it is also possible that the medium of communication may be selected for symbolic meaning that transcends the explicit message. In this way, the medium itself becomes a message' (p.84). This approach to the new media fits perfectly into our model of the social context as symbolic order, practically implemented and embodied in artifacts. If we consider the general impact of CMC on organizations and individuals, we see the decay of old communication models and the emergence of new ones.

First, we see that the model of communication as the passage of information from one person to another is becoming obsolete. This model, which stood up perfectly well for decades, is now in a state of crisis, partly because of some of the peculiar features of electronic environments, such as the asymmetry between message sender and message receiver; this takes on a special physiognomy in CMC, as we shall see later. The old model, disrespectfully called the 'parcel-post model' by Tatar *et al.* (1991, p.185), is criticized for conceiving communication in an entirely inadequate way, as the transportation of an inert material – the information that actors exchange with each other – from one point to another along a 'pipeline'.

The most important thing, according to this model, is that there should be no cracks in the pipeline through which information can leak out. The model of communication as information transfer does not take into account the cooperative component, which stimulates reciprocal responsibility for successful interaction and a series of subtle adaptations among interlocutors. It does not consider the fact that it is possible to communicate only to the extent that participants have some common ground for shared beliefs, recognize reciprocal expectations and accept rules for interaction which serve as necessary anchors in the development of conversation (Clark and Shaefer, 1989). Dohény-Farina (1991) observes: 'The theory of communication as information transfer separates knowledge from communication; it treats knowledge as an object that exists independently of the participants in the innovation venture. With this independent existence, information becomes an object that can be carried through channels' (p.8).

The new, alternative concept, which is emerging with increasing clarity, is that communication is a common construction of meanings (Kraut and Streeter, 1995); our conceptualization of the social context supplies this view with a fitting theoretical framework. The old model is based on an essentially incorrect idea of what knowledge really is. It assumes that: 'A body of information, of objective facts, is just lying out there waiting to be communicated. When the communication is successful, the receiver is put in possession of those facts. The facts determine the communication, unless the originator interferes. The job of the originator is to move the facts from one place to another, handling them as little as possible so as not to tarnish them' (Dobrin, 1989, p.60). Our approach (in particular, the sections on the interpretation of situations) tells us that to consider information as an inert material to be transferred, as if it were a load of gravel being transported by barge downstream, is quite misleading. Information, like communication, is always a means to an end. It is significant and strategic in that it is produced and used by actors to attain their goals in the daily 'funny soccer game'.

Second, we see the rise of a new model of communication as a global connection, which is captured by the network paradigm. The development of electronic networks has accelerated the still ongoing integration between the communicational and organizational dimensions of social processes. This constitutes the theoretical and practical drive of the network paradigm (Rice and Love, 1987; Rice and Shook, 1990), which considers communication no longer as the transfer of a message from a sender to a receiver, with the return of the information to the sender closing the cycle, as in Shannon's classic linear model, but as a process taking place among a set of actors inside a developing relationship (Stasser, 1992). By 'refocusing attention away from individuals as independent senders and receivers of messages, towards individuals as actors in a network consisting of independent relationships embedded in organizational and social structures' (Rice, 1990, p.629), the emerging network paradigm seems to be highly compatible with our model of communication.

The network paradigm does not apply only to individuals. The boxes of the matrix representing the network may be groups, organizations or institutions. What is important is that they are not considered as isolated bi–univocal relationships, but as a complex web of interactions among all participants. Relationships are represented on an $N \times N$ grid, N being the number of participants. Single boxes (i, j) contain values which indicate the intensity of the link between actor i and actor j, an intensity which defines the presence, or the force, or the frequency of the ongoing communication, or any other aspect in which we are interested. The network paradigm offers a unified approach to communication and organization (Mansell, 1993), which are no longer considered as exclusively individual processes. Thinking in terms of networks involves taking into account the context in which actors move, and suggests that communication is part of an overall relational grid.

The main drawback to the network paradigm has been identified by some social psychologists in the 'system rationalism' often connected to it. Lea (1991) uses this expression to designate the markedly managerial standpoint that has inspired studies on CMC like those of Caswell (1988), Hiltz (1988) and Rice (1986). According to Lea, 'system rationalism' attributes excessive importance to the efficiency of the new media while underrating the importance that social communication processes have in CMC. A similar remark may be applied to studies which, like that of Malone and Rockart (1991), stress the reduction of costs and the speed, reliability and efficiency of message transmission in CMC, while neglecting its more properly social aspects.

The novelty of electronic communication environments cannot be reduced exclusively to possible economic advantages measured in terms of the balance between its costs (financial or human) and benefits (productivity increases). Adaptive structuration theory (AST) (see below) also refutes the idea that the new technologies, as such, have a certain univocal impact on organizations. This idea arises from the assumption that a unidirectional causal relation exists between technologies and social contexts – but this is precisely what AST denies. It is a mistake to assume that the impact of technologies on individuals, groups and organizations is generally beneficial for them. Technological innovation in itself and for itself does not resolve the problems of individuals, groups or organizations. On the contrary, we often find ourselves faced with spiny problems precisely because we have to use those new technologies.

The question: how do new technologies change people, groups and organizations? only deals with half of the problem. It should be completed with the second half: how do people, groups and organizations modify and adapt the new technologies to suit them? We have to deal here not with unidirectional linear causality, according to which new communication technologies are supposed to act on people by necessarily improving their performances, but with a subtler, situated and usually circular influence. Every time the peculiar technical components of a given socio-technical system interact with some of its social components, we may ask ourselves which kind of influence processes we could expect on this particular occasion.

In a wider evolutionary perspective, there is no doubt that human societies generate new technologies, not vice versa. Our species has the adaptive capability to invent and use many sorts of artifacts (Cole, 1990, 1995), including cognitive ones which, alone, can be neither created nor adapted. Artifacts can fit their intended environments to the extent to which human beings adapt them and – especially – adapt themselves to those artifacts (Collins, 1990) to make them work as proper tools.

Third, in the wake of criticism of 'system rationalism' we note the appearance of another research line opposing technological determinism. This is adaptive structuration theory (AST; Contractor and Eisenberg, 1990; Contractor and Seibold, 1993; Poole and DeSanctis, 1990, 1992), according to which the effects of new communication technologies emerge from the specific set of complex social interactions existing between users. Rejecting the tacit assumption that technologies have general, necessary and uniform effects on people's living and working environments, AST examines actors' everyday practices to see what uses are effectively made of technological innovations. AST attempts to explain how it is that groups of similar composition working on identical tasks can perceive and use the same technological resources in very different ways.

This new perspective has been developed by Poole and DeSanctis (1992) and processes suggestions coming from a more distant relative, Giddens' (1984) structuration theory. AST can also legitimately boast of other ancestors such as the socio-technical approach of the Tavistock Institute in London (Trist, 1981; Rousseau, 1983) and the theory of structural contingency, which inspired organizational studies for many years (Woodward, 1965; Perrow, 1970; Gutek, 1990). The basic working hypothesis of these theories is that relations between technological innovation on the one hand, and the functioning of organizations on the other are the result of reciprocal adaptation which takes on different forms according to different social contexts. For example, Walther (1994) shows that CMC is more readily accepted when it is used to connect people who have to work together for a certain period of time than when it is used to produce 'one-shot' groups.

7.2 CMC AND ORGANIZATIONS: THE MYTH OF ELECTRONIC DEMOCRACY

The new technologies offer an important opportunity to bind organization and communication closely together. On the practical plane, they promote progressive identification between the structure and functions of organizations with the structure and functions of the computer networks conveying the information which is conceived as the main resource for organizational survival. On the symbolic plane, they strengthen the power of the network metaphor, so that it captures at the same time both the corporate communication system, understood as the existing set of computer connections, and the company organizational system, understood as the web of actors' mutual commitments and strategic interactions. We see that technological artifacts exert a double influence. On the one hand, computer artifacts constitute the physical aspect of the interface mediating between people, between people and organizations, and between one organization and others. On the other, they supply the set of metaphors which are used to explain social and cognitive processes taking place within and between actors. These social and cognitive processes are still frequently expressed in the language of information processing and information transfer. The models of knowledge as information processing and of communication as information transfer still remain alive in the commonsense of both designers and users, and will remain at least until their crisis is fully over and they have been solidly replaced by the emerging alternative model of communication of construction of a shared meaning.

It is within the paradigm of information transfer that the idea arises that CMC, by its very nature, favours communication and that it may therefore contribute efficiently to the democratic development of organizations. The implicit assumption is: the more information is available to single actors, the more democracy is possible in organizations – democracy being understood as the possibility for actors to participate in decision-making. Sproull and Kiesler (1991) explain this position clearly: 'In a democracy, people believe that everyone should be included on equal terms in communication; no one should be excluded from the free exchange of information. Independent decision-makers expressing themselves lead to more minds contributing to problem-solving and innovation. New communication technology is surprisingly consistent with Western images of democracy' (p.13).

This view seems questionable for three reasons. First, the assumption that CMC favours social and political development in a democratic sense simply by virtue of its introduction into an organizational environment rests on technological determinism criticized above. Second, it presupposes that CMC functions uniformly in all environments, indifferent to the organizational contexts in which it is used, unlike the tenets of adaptive structuration theory and of the socio-technical approach. The third reason why we are led to question the presumption of the intrinsically democratic nature of CMC is that it depends on an inadequate idea of the relationship between information and knowledge, according to which greater amounts of information automatically lead to greater knowledge and greater freedom in decision-making. In our view, it is situated knowledge (see above), not mere information, which is crucial for the development of actors' responsibility and participation in decision-making.

Let us examine more closely the solidity of the assumption that CMC is capable of overcoming social barriers in organizations. This idea, originally proposed by Hiltz and Turoff (1978), has now reached us through the works of Feldman (1987) and (mainly experimental) studies of social psychologists working with Sproull and Kiesler. Their results do not appear to be confirmed by research – this time in the field, involving ample sets of users in various application sectors – carried out by the Institute for Research on Interactive Systems of the Rand Corporation (Bikson, 1987; Bikson and Eveland, 1986, 1990; Eveland and Bikson, 1987) over quite a long period of time (going back in fact to 1982). Bikson et al. (1989) have shown that not only is CMC not particularly suitable for lowering social and hierarchical barriers in organizations, but that it generally tends to strengthen existing patterns of interaction instead of creating new ones. The research of Saunders et al. (1994) on the use of teleconferences in health care environments confirms the persistence and sometimes the reinforcement of status barriers.

To expect technologies to solve social problems in organizations would be imprudent, to say the least, seeing that the assimilation of new technologies requires proper use of organizational resources, not the opposite. The capacity of organizations to profit by the potential benefits of CMC 'heavily depends on the resolution of social questions about collaboration – questions about group norms and values, equitable role structuring, and shared task management – that organizations introducing new technology are not usually prepared to address' (Bikson et al., 1989, p.90). Spatial barriers too, which CMC easily overcomes, are very often far less permeable than we might expect. Of all the messages among the non-technical staff of the Rand Corporation, excluding coast-to-coast communications (which offer particular advantages from the viewpoint of the temporal organization of cooperation), 45% were sent to people who were actually physically very close (Bikson et al., 1989).

Clearly, one key element in using CMC is the type of organization containing the network: it defines what kind of people use it, their goals and professionality, and current company policies regarding communications. So it really comes as no surprise to learn that, in a company with a sophisticated organizational structure like IBM, no less than 83% of e-mail messages are sent within single divisions, and that 93% are addressed only one hierarchical step higher or lower than that of the sender (Smith et al., 1988). Rice (1990) also estimates that CMC increases rather than reduces status differences in most organizational contexts.

These results are congruent with the working hypotheses of the socio-technical theory, according to which the use of technologies is modelled by social and organizational contexts. From a different but complementary viewpoint, not regarding differences of level within organizations but gender differences, women's studies are also quite critical of the fact that the new technologies necessarily operate favourably towards employees – particularly women – who occupy the lower levels in organizations (Green et al., 1991).

A second argument brought forward by Sproull and Keisler (1991) in favour of their thesis is that CMC allows all members of an organization equal access and equal participation in communication, thus favouring low-status employees who are finally free to express themselves without being penalized. However, we believe that equitable access to the network cannot really be sustained as a general criterion, for at least two reasons. First, electronic environments often present considerable problems of usability (van der Veer et al., 1990). Due to the inadequacy of some interfaces, inexperienced or occasional computer users may not even know whether a message is really winging its way or not. A second reason why access to networks is limited is that 'CMC systems are, and will be for a long time, too inaccessible – physically, culturally, technically, and economically – for most people. Thus, CMC systems represent another potential source of social inequity' (Rice, 1990, p.642).

In fact, we must expect a phase of bitter struggle for power accompanying the introduction of new technologies. Strategic information manipulation is inherent to organizational life. Actors seek to achieve their interests by both representing and misrepresenting information, and the new technologies offer some professionally competent groups excellent opportunities for making strategic use of both access to and control of information. Zmud (1990) observes: 'Traditionally, an organization member's zone of information influence has been limited by a number of constraints, most of which reflect task design, authority relationships, and physical, geographic, and temporal boundaries. New information technologies relax many, if not most, of these constraints. ... As a consequence, the incidence of strategic behavior will increase, precisely because of this enlarging of individuals' zone of information influence and an increasing willingness of many of an organization's members to delegate important information processing behaviors' (p.114).

Delegation to technologies favours particular socio-professional groups – precisely those groups that are in a position to gain the greatest benefits from the possible increase in the strategic use of information. When new technologies are introduced in an organization, the presence of different levels of technological competence among its members tends to modify the pre-existing distribution of power so that technically competent members acquire greater power. We believe that the new communication technologies do not generally favour democracy in organizations, except in the sense that they create occasions for redistribution of influence and authority in favour of new groups (McGrath, 1990). They might even erect new barriers of competences and alliances between actors and socio-professional groups, above and next to the old barriers.

Sproull and Keisler (1991) consider that CMC promotes equity among group members, not only in acceding to the network but also in making contributions to communication. In a series of experiments with students and recent graduates, they found that, in CMC, each group member tended to speak for an 'appropriate' length of time, i.e. about one-third of the time available for group discussion (experimental groups being composed of three subjects each). As it is well-known that higher-status group members tend to speak for longer in face-to-face inter-actions, Sproull and Keisler concluded that, in electronic communication, with respect to similar face-to-face communications, higher-status members reduce their interventions while lower-status members increase them.

Dubrovsky *et al.* (1991), again using students in an experimental environment, confirmed this effect, and added a series of observations on the first interven-tion in the discussion. While in face-to-face communication, the first interven-tion – which conditions the subsequent discussion and strongly influences the decisions taken by the group – is usually reserved for high-status members, in CMC first interventions are more equally distributed among group members. However, we must note that, in CMC, unless special rules are followed, there is no real turn-taking. Everyone can write their messages on their keyboards when-ever they want, without having to contend directly with other participants to gain audience. In CMC, we can hardly think of any real 'first' intervention, in the sense that this term has in a social context involving actors' physical co-presence.

Results on equity in participation were also questioned by Adrianson and Hjelmquist (1991) in a field research on the daily activities of workers at the Swedish Research Institute for National Defence. No equity effect in terms of time taken by various group members in CMC was found. The authors explain the discrepancy between their results and those in favour of equity in participa-tion by stating that most previous research had been carried out in the labora-tory with university students or graduates, in situations of dubious ecological validity.

The question of type of experiment, often oversimplified with respect to the everyday contexts which it aims to explain, is a serious problem in CMC research. We may feel slightly uneasy while reading about the effects of CMC on status differences among members of a group, in which those status differences consist of the fact that a few students are one or two years ahead of the others and are therefore classified as high-status, whereas others further behind are low-status. We must not forget that, in real life, groups and organizations are characterized by very particular contexts and meaningful histories (Fulk *et al.*, 1992), and that experimental research that neglects these essential dimensions may be quite impoverished.

So there does not appear to be any clearly observed proof of a generalized effect of CMC in favour of democratic growth in work environments. On the contrary, Austin *et al.* (1993) have shown that, in poorly sophisticated commun-ication systems for group decision-making, i.e. when the system does not have the task of distributing participants' interventions according to pre-established methods, control of the interaction is generally far from equitable. They found that the more capable group members, especially those more competent than others in using the new technologies, tend to monopolize the space available for interaction. This is particularly so when the group contains members who are inexperienced in computerized technologies, who are always heavily penalized.

It is also necessary to bear in mind the fact that power relations are not only top-down. Even communications between people of equal level convey power relations. Peer group pressure, usually important as a source of influence, is particularly felt in moments of uncertainty, e.g. when companies start cutting down on staffing levels (Klein and Kraft, 1994). In general, conflict does not only occur top-down, between employees and bosses and vice versa; it permeates the entire system of relationships within the organization. Conflict may involve actors of equal level and status and also actors belonging to the same socio-professional groups, as long as the situation requires it on the basis of competition for the control of strategic resources.

To presume that new technologies are equipped with the intrinsic capacity to transform organizations in a democratic sense is the result of a twofold flaw: technological determinism, and the adoption of an inadequate model of communication as the simple transfer of information. If we accept the interpretative framework of our model of social context, we may consider the media as equipped with peculiar symbolic significance, and not simply as channels for the passage of information which is presumed to be neutral. Our perspective shows information, knowledge and communication as constantly bound together in any act endowed with a socially recognizable meaning.

Social contexts assign definite meaning to the use of technologies, which in themselves are directed neither towards democracy nor towards authoritarianism. 'Information Technology (IT) is best viewed as neutral regarding its organizational consequences. The nature of an organization's IT use emerges through complex interactions among the intentions of key actors, attributes of the technology involved, and dynamic organizational processes' (Zmud, 1990, p.95).

7.3   ARE INDIVIDUALS TRULY ISOLATED IN THE NEW ENVIRONMENTS?

The same perspective that gives rise to the idea that CMC can promote democracy in organizations ascribes to it univocal and beneficial effects even on an individual scale. According to this approach, people are more free to express themselves in electronic environments, although the other side of the coin is that they may sometimes express themselves in a more impulsive, rude, or even irresponsible way. Both outcomes, one socially positive and the other negative, are credited to the fact that, unlike face-to-face relationships, social rules are weak or absent in electronic environments. People consequently cease to be inhibited by considerations of status, either their own or that of other people, and by the fear of being criticized.

We agree that in special cases (e.g. electronic interviews and some systems of consultancy for sexual problems) CMC does allow people to be more open and able to give free rein to their self-expression because they feel protected by the relative anonymity of CMC and because they feel (mistakenly) that CMC is 'ephemeral' in character (Sproull and Kiesler, 1991). But whether electronic environments do in general make people feel less inhibited is highly questionable. Valacich *et al.* (1994), for example, attribute the good results of electronic groups in generating new ideas (found in experimental conditions) not to any reduction of social pressure, but to better temporal organization of the passage connecting the emergence of an idea to its expression. CMC seems to reduce the 'waiting time' between the generation and expression of ideas and thus to miminize the effects of the 'production block' which in ordinary face-to-face conversations deters participants from expressing their ideas immediately as soon as they come to mind in order to avoid interrupting the speaker. While one person is waiting his turn, he forgets or loses interest in his still unexpressed intuitions.

CMC is considered by Sproull and Kiesler (1991) as capable of letting people express themselves more openly because it isolates them from social contexts: 'People interacting on a computer are isolated from social cues and feel safe from surveillance and criticism. This feeling of privacy makes them feel less inhibited with others. It also makes it easy for them to disagree with, confront, or take exception to others' opinions' (pp.48–49). People who are shy, who feel guilty or disturbed by conflict with others, may find protection in the presumed isolation of CMC. We believe that, for those with relational difficulties, the apparent isolation of CMC is only one aspect of the complex dynamics governing communications in groups and organizations. In any case, it seems to be misleading to attribute a generally repressive meaning to social contexts. On the contrary, as we know, social networks often afford a great deal of emotional and practical support to those in difficulty.

Assuming that CMC operates in a social vacuum, electronic communication appears as an environment in which the personal identity of the sender and of possible receivers tends to become blurred. That is, a condition of de-identification occurs, in which people lose their sense of personal responsibility and of the respect due to social norms. Proof of this comes from the controversial phenomenon of 'flaming' (Siegel *et al.*, 1986), i.e. rude, impulsive behaviour, and the use of crude or even insulting expressions, which some authors consider typical of electronic communication environments. The term has now entered common usage, and episodes of intolerance expressed through CMC, usually denoting rage and frustration, are currently called 'flames'.

Both the existence and the interpretation of 'flaming' in terms of de-identification produced by the lack of social cues have recently been questioned. Lea *et al.* (1992), reinterpreting data from previous researches, state that flaming is relatively rare and usually caused by 'frustration arising from the communicational inefficiency of the medium' (p.98), especially when the person concerned is under stress, rather than by any lack of proper social and normative contexts in CMC. Another explanation, complementary to that of Lea and co-workers, views 'flaming' as an expression of the culture of American university students. Studying electronic communications among students on an American campus, Weedman (1991) noted that the formulas used in 'flaming' may be regarded as part of the students' culture and that the gross language used generally has often a playful sense and serves to transmit positive emotions. That is, recourse to 'flaming' aims at strengthening group identity and has no antisocial sense.

CMC is expected to stimulate the development of impulsive behaviour not only in the way of expressing oneself, but also in decision-making: Siegel *et al.* (1986) ascribe to it the capacity to cause marked group polarization, i.e. to stimulate people to make extreme decisions. According to Sproull and Kiesler (1991) too, CMC makes people act in an impulsive and anti-normative way, by removing the most important contextual elements from the communicative situation. These data and their above-mentioned interpretation were questioned by Lea and Spears (1991) and Spears and Lea (1992), after careful examination of available empirical results and applying concepts from social psychology studies. They refute both the idea that group norms are absent in CMC and the related view that those who communicate by CMC are anonymous and socially isolated. 'Empirically, the assumption that CMC is characterized by a weakening of social norms seems to have little direct or independent support. In fact, it could be argued that an absence of social cues from other interacting individuals, together with the resulting uncertainty, forces people to resort to default social norms to guide their behaviour' (Lea and Spears, 1991, p.286). The framing of each particular situation depends on the characteristics of the context in which the electronic communication develops (Spears *et al.*, 1990): when the context enhances a person's social identity as a member of a group, i.e. makes it particularly conspicuous and important for that person, then he or she is particularly sensitive to the norms of that group and behaves accordingly.

Discussion about the presence of social norms in CMC reflects discrepancies in the way of conceiving of people, as basically isolated individuals or as social actors capable of constructing social identities for themselves. Sproull and Kiesler see people communicating by computer as essentially isolated. Their work, *Connections*, opens with the question: 'How do people treat one another when their only connection is a computer message?' (p.ix), which reflects their peculiar approach to CMC. In reality, people are never connected 'only' by means of a computer message: they are linked by an organizational network that parallels the technological one and also by a set of partly shared expectations, needs and goals, which are to some extent reciprocally recognized. Moreover, CMC generally does not work in isolation as an alternative to other media, but fits into the pre-existing media with which it constantly interacts (Kling, 1980, 1994, 1995; Kling and Scacchi, 1982).

In contrast to this rather asocial vision of CMC, researchers supporting the theory of social identity believe that individuals carry the social world and its rules within themselves (Tajfel and Turner, 1986; Hogg and Abrams, 1988). They believe that people have unrepeatable identities as unique individuals and that they also have social identities as members of one or more social groups, real or fictional. Belonging to one or more reference groups is part of an actor's identity. Contexts, according to the relationships created between actors' situated interests and the specific situations they face, emphasize the social or individual poles of personal identity.

As always, situations have to be interpreted: Mary must decide whether John's complaints about her lukewarm interest in cooking refer to her as a person (daughter of her mother Ruth, who is an excellent cook, etc.), as a wife (a member of the category – ideally very large – of married women), or as a woman (in which case Mary is a member of a far larger category). Obviously, daily life contains multiple competing categorizations regarding social identity: for example, Mary may be a young career woman living in New York but originally come from New Orleans, and each of these connotations may activate particular aspects of her social identity which may be made salient by the situation, i.e. John complaining about her lack of interest in *haute cuisine*.

It is because of categorization and identification processes that Lea and Spears believe that CMC does not necessarily involve isolation, de-individuation, and lack of social norms. They reformulate the concept of de-individuation in the sense that anonymity, if associated with immersion in a group, does not weaken social norms but the opposite, increasing the saliency of the group and its norms (Lea and Spears, 1991; Spears and Lea, 1992). Therefore, whether CMC makes the social identity of the participants more or less salient, with the consequent greater or lesser activation of group norms, depends on the type of pre-existing group and the tenor of the ongoing communication. The social world is not only outside but also inside people, as part of their identity, and functions even when they sit – physically alone – in front of their computer screens.

Here too, as noted in the previous sections, the relation between CMC and actors may be viewed from two quite different standpoints. One view, that of Sproull and Kiesler, is that CMC may be conceived as a process of influence starting from technology, developing in a univocal, generalized and unidirectional way. This is the view of those who accept the idea that CMC as such involves anonymity, de-individuation and loss of social norms. The other view – our view, which is closer to that of Spears and Lea – is that the relation between CMC and actors may be seen as multiple, context-dependent and circular. In this case, the experience of those who communicate by CMC depends on the type of group and the type of context.

7.4 ASYMMETRIES, RULES AND PRIVACY IN ELECTRONIC ENVIRONMENTS

The relationship between the senders and receivers of messages in CMC has no exact counterpart in face-to-face communication. The latter comes into being and develops in a cooperative framework actively managed by its participants (Goodwin and Heritage, 1990); the same cannot generally be said for CMC. Face-to-face conversation takes place between two or more people who, as well as being physically co-present, agree to interact and negotiate the forms and contents of their interaction in the light of their goals. In face-to-face conversation, cooperation between interlocutors is constantly monitored by a series of fine adaptations, turn-taking and reciprocal corrections. Even the person who is not talking is not passive, but is continually called upon to contribute – usually paralinguistically – to the development of the conversation, confirming the correct reception of utterances, interrupting their flow with requests for more details or making facial expressions of agreement or disagreement. In situations when interlocutors are co-present, messages are processed in real time, with no delay between emission and reception. Each message has to be understood and accepted, at least provisionally, for the communication to continue (Clark and Wilkes-Gibbs, 1986; Clark and Schaefer, 1986; Clark and Brennan, 1991).

Instead, in CMC, the collaborative framework is very weak (McGrath, 1990; Brennan, 1991); it lacks the possibility of real-time confirmation of its participants' intention to collaborate, cooperation during message formulation, and immediate processing of messages (the latter phase includes negotiation of the social significance of the messages and clarification of how they are to be interpreted). Even the most elementary forms of feedback are unusual in CMC: 'No one sends a message to "just" say ⟨yes⟩ or ⟨that's right⟩. One possible explanation is that the Unix community strongly discourages users from sending ⟨me too⟩ messages that take up valuable disk space' (Fafchamps et al., 1991, p.219). And yet it is precisely this type of message, only apparently banal, which actors exploit to request and obtain support and consent from other people when they are engaged in face-to-face communication.

CMC is sensitive to organizational climates, although this opinion may not be shared by those who think that CMC is by nature capable of promoting democracy and cooperation. In fact, low-status people can send their messages to anyone connected to the network and may also be reasonably sure of their arrival, but they do not know if they have actually been received, i.e. read, by the person to whom they were addressed. In CMC, senders cannot know whether their messages have reached their destination until they receive a direct or at least an indirect reply – something which is not usual in many electronic communities and for several classes of messages. In this sense, electronic communication is a strange environment: the transmission of information may be systematically separated from its reception.

One characteristic of CMC as a technological artifact is precisely that of facilitating information transfer without at the same time establishing a normative framework which ensures, or at least favours, message reception. CMC thus takes on the physiognomy of an environment of *virtual* communication. On the one hand, it poses some technical premises for the development of possible cooperative processes and related strategic games by setting up lines of interconnection among actors; on the other, it does not define the rules for its proper use. Electronic networks tend to overwhelm actors with messages without providing them with the social bonds which justify the costs of communication and cooperation. Faced with the increasing numbers of undesired and often irrelevant messages flooding in, receivers may be obliged to defend themselves by refusing whole classes of messages which are considered *a priori* insignificant or at least of low priority.

Electronic networks are plagued by a phenomenon that happens to practically everyone at certain times of the year, e.g. before Christmas, when our mailboxes are crammed full with (for us) low-class communications from companies offering everything under the sun, from cut-price trips to the Caribbean, to jewellery and silver plate, and hampers of gastronomical delights and fine wines. Those who are not interested throw everything away without even looking at it. Have the sales networks really got in touch with the people to whom they wanted to sell? we could say *yes*, if those who open the envelopes actually examine what they contain, but definitely *no*, if people just throw them away without even looking to see what is inside (although we nearly always know what their contents are likely to be).

Senders not only cannot know *a priori* who will open their envelopes, but generally do not even know *a posteriori* either, unless a response comes in the form of an order or a phone call asking for more information. But they do know that, the greater the number of special offers, the greater the risk that receivers will *a priori* refuse even to examine them. Top managers in organizations must sometimes choose whether to spend their entire day reading their e-mail or whether to have it filtered before it reaches them by telling the system, for instance: 'No incoming mail under vice-president level'. Only a small percentage of managers admit to using filters, but complaints about the indiscriminate proliferation of electronic messages are quite widespread among e-mail users (Arensburger and Rosenfeld, 1995; Maes, 1994).

In electronic environments, there is an asymmetry in the message–sender/message–receiver relationship which has no counterpart either in face-to-face or in telephone or postal communications. This is because – apart from the wearisome phase of keying in data – sending a message in CMC involves costs that are far lower, in terms of both time and money, than those required by other forms of communication (with the possible exception of printed publicity, which also poses problems of overload, irrelevance, and the need for filtering by receivers). For senders in CMC, in terms of effort or expense, the difference between sending a message to a single person, a group, or the whole network may be almost nil. The overall result is that CMC, thanks to its efficiency and economy, tends to produce an abnormal mass of messages, which often exceeds the receivers' capacity to process them.

The solution to this problem, or at least the not over-conflictual management of the situation, is referred in CMC to the possible development of social rules. In fact, organizations already have social rules, although they are often only implicit. They are the same rules that model the communications flowing through the network and which pose those restrictions we found in studies on addressing messages in companies such as IBM or Rand. The question of social rules in CMC has also turned out to be a thorny issue in the development of teleworking environments.

Research on teleworking considers two or more work environments, maybe thousands of miles away from each other, linked by computer networks as efficiently as possible. The 'Portland experiment', carried out by the Palo Alto Research Center (PARC) of the Xerox Corporation from October 1985 to January 1988, connected the PARC Systems Concepts Laboratory (SCL) in California to a similar work group in Oregon. The two units had to develop activities to be undertaken together in a common work area. The unit in the SCL had 13–15 researchers and the one in Oregon 8–10, together with other technical and administrative personnel, plus visiting experts, consultants, etc. The aim of the experiment was to develop communication systems that would increase interdependence among people and between the two units. The priority aim of the experiment became that of making the two different, separate environments as similar to a single environment as possible (Olson and Bly, 1991). However, this aim, typical of the most highly structured forms of telework, came up against a series of limitations, both regarding the technical capacities of the new media to make the reciprocal presence of actors as evident as possible, and regarding cultural rules on the use and destination of space – private spaces being no longer, in CMC, clearly marked and socially recognized as they used to be in traditional environments.

The threat to actors' privacy inherent in the shared spaces linked by video and sometimes also by audio, together with message systems like e-mail, should not be underrated (Clement, 1994). It has created a new source of worry on the ethical plane: 'A person should always be able to know if someone else was looking at them; such a feature was built into an application for controlling the video/audio connections'. Concern about *ethical video* begins to emerge through statements like: 'The observed should always be able to observe the observer' (*ibid.*, p.225). We realize that reciprocal contact cannot be imposed on people. But is the acknowledgement of the need for permission for reciprocal access an appropriate response to the challenge that the intrusiveness of the new communications technologies poses to actors' autonomy?

The fact is that, in the new media, the distinction between private and public spaces becomes unclear. There are no longer any private spaces in a system in which a video channel, kept permanently albeit consensually open, links two offices 24 hours a day, as happens at the EuroPARC in Cambridge (Heath and Luff, 1993). A threat to privacy is a danger, not only for people's independence but also for the development and preservation of their identity. Privacy is not simply a mechanism which allows people to regulate access to themselves in a flexible way; it is also an important instrument defining the limits and boundaries of the self. When the permeability of these boundaries is regulated by individuals with respect to their current needs, firm perception of individual independence may develop (Altman, 1979). But we see that perception of control over actors' interactions with the environment may be placed in a state of potential crisis by the invasive capacity of electronic environments. Their present philosophy, which presents the widest possible access to information and its transmission to the largest available audience as valuable goals to be achieved, may lead to reduced privacy and to the sacrifice of some of the social norms which almost automatically protect it in face-to-face communication.

However, this does not mean that the social and normative context is absent in CMC, as Sproull and Keisler (1991) and other researchers supporting social presence theory believe. They usually compare CMC with face-to-face conversation in order to stress how the former bears almost no traces of interpersonal social cues. Actually, CMC does lack many of the cues typical of interpersonal aspects of the social exchange, which are so extensively present in face-to-face conversation. The point is that the view of social context which inspires this comparison and is implicit in social presence theory is not only unacceptable in general, because it reduces the social world merely to direct physical connection between people; it is also singularly unsuitable for discovering where the social dimension of CMC may be found. It is true that, if we start from an idea of social and contextual as equal to relational and interpersonal, we have to admit that CMC is very poor in social cues, because it does not transmit the information which face-to-face communication instantly provides about other people. CMC cannot convey the tone of voice of interlocutors, their pauses, facial expressions, etc., or can only do so to a very limited extent. Since the physical presence of others – with its related wealth of information on the psychophysical and emotional state of the interlocutors – is lacking in CMC, Sproull and Kiesler think that electronic communication takes place in a sort of social vacuum.

Instead, Spears and Lea (1992) assert that the social world in CMC does exist, provided that we know where to find it, i.e. in symbolic processes. These authors follow the theories of social identity of Tajfel and Turner (1986) and of self-categorization of Turner et al. (1987). Social context is present in CMC because categorization and identification processes – of the self, of the other, of one's reference group, organization and situation – all function powerfully within it. In our perspective, these processes play a central role in structuring social contexts. They make possible interpersonal relationships: people communicate to the extent in which they live in common symbolic systems.

In certain conditions, the poverty of the interpersonal relational framework in CMC may highlight the social structure of the context resulting from processes of categorization, belonging, identification. According to Spears and Lea (1992), and in agreement with our model of context, social signals may function even more powerfully on the basis of pure symbols (of power, status, authority, etc.) when they are not circumscribed and mitigated by interpersonal cues such as a friendly expression, an irritated tone of voice, or a casual gesture of tiredness. Consequently, the influence of social context may be even stronger in CMC than in face-to-face conversation. A study by Weisband (1994) offers experimental proof of this fact. It depends, each time, on which aspects of personal and social identity are made salient in the various situations (Lea and Spears, 1991). Spears and Lea (1992) openly challenge the idea – inherent in social presence theory – that social context may simply be reduced to the physical presence of other actors: 'The social identity approach rejects the individualistic meta-theory of interdependence, viewing social categorization and social identification as the critical mediators of group processes' (p.54). In their view, which we accept, 'The group is conceptualized in socio-cognitive, and not first and foremost in structural or relational, terms. This then allows for the impact of group influence independently of the co-presence of others, and also for the influence of more general social categories defined independently of individuals *per se*' (*ibid.*).

It is not enough to be physically alone in a room to be removed from the social world. Social context is not something outside or above people; it is both around and inside them, an integral part of their identity. Contexts are interwoven with principles and symbols, by means of which they inspire people's actions and are in turn reshaped by their practices. In this very sense they might be considered as 'portable contexts' (Brown and Duguid, 1994), capable of 'transmitting authority' and also of 'sustaining interpretations' that actors give to situations. We can realize this fact only if we cease contrasting individuals and social context. Some research approaches in social psychology lead us to persist in regarding this opposition as a matter of fact: 'researchers are subtly encouraged to draw relatively rigid conceptual boundaries between the individual and the social context. Social context thus becomes something outside the individual – something that is definitionally independent of individuals but which has penetrating effects on their behaviour' (Fulk *et al.*, 1992, p.9).

Context, especially at its cultural–normative level, is critical for the success of new technology-mediated communication tools. Kraut *et al.* (1994) investigated why, of two video telephone systems very similar in structure, one was accepted and prospered while the other was rejected and died. They found that taking into account the fit between the tasks to be accomplished and the characteristics of the systems – as required by the current utility approach – was not enough to explain the different outcomes. Social influence processes also had to be considered: 'The importance of work group went beyond providing a critical mass of communication partners. The work group also provided a safe environment to try out the new systems, and through examples and discussion illustrated norms about how the system should and could be used' (p.20). Actually, the cost–benefit approach and social influence perspective are not in contrast; in fact, they must be combined if we wish to understand which attitudes potential users will develop towards CMC systems.

Groups facing electronic communication environments use time and resources to invent, test and make mutually recognized new norms, principles and styles of proper interaction, which must also include rules of etiquette covering sensitive matters like intrusiveness and privacy. This is why people need a human interface when starting to use CMC, as Fulk (1993) found in her study of e-mail use among a group of scientists and engineers working in the production research company of a major petrochemical corporation in the United States. A human interface is necessary because 'formal or informal peer training effectively uses social influence processes' (p.945). Relying on co-workers for help means asking not only for practical guidance but also for normative directions defining the proper use of the medium.

CMC environments do not have a special meaning by themselves, as mere technological objects. They receive their situated meaning from their social context: 'Symbolic features need not be fixed attributes of a medium. The symbolic meaning may well arise, be sustained, and evolve through ongoing processes of joint sensemaking within social systems' (Fulk, 1993, p.922). The symbolic significance which actors attribute to the various forms of CMC is no less important, in choosing and using the media, than considerations regarding their cost and reliability.

Normative social influences mingle with informational influences, as Figure 4 shows, in determining the meanings assigned to the new media and the type of messages which may appropriately be channelled through them. Researches such as those of Lea (1991), Trevino *et al.* (1990), Fulk (1993), and Fulk *et al.* (1992) show how recourse to the analytical categories of symbolic interactionism may be profitably used to study the everyday use of the new media.

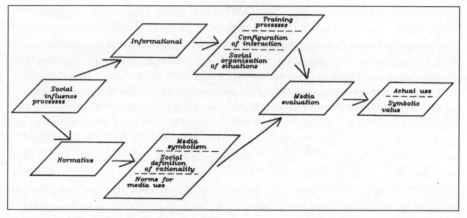

Figure 4. Social influence processes involved in the choice and use of new media.

CHAPTER EIGHT

Virtual Environments and the Development of Possible Selves

Individuals find friends and groups find shared identities on-line, through the aggregated networks of relationships and commitments that make any community possible. But are relationships and commitments as we know them even possible in a place where identities are fluid? ... In cyberspace, everybody is in the dark. We can only exchange words with each other – no glances or shrugs or ironic smiles. Even the nuances of voice and intonation are stripped away. On the top of technology-imposed constraints, we who populate cyberspace deliberately experiment with fracturing traditional notions of identity by living as multiple simultaneous personae in different virtual neighborhoods. (Rheingold, 1993)

As aspects of the virtual environment become part of our natural environment, the distinction between computer reality and, for lack of a better word, conventional reality will become increasingly blurred. (Shapiro and McDonald, 1992)

8.1   AN ALTERNATIVE SPACE FOR COMMUNICATION AND BELONGING

Virtual reality (VR) is now placed at the front of the stage where it is brightly lit by the floodlights of the mass media, more for what it promises – the creation of alternative worlds, disembodied and infinitely malleable – than for what it effectively provides, at least at the present time. Its applications are in fact still quite far from allowing experiences as realistic as those which are available in daily environments (although it must be admitted that great strides have recently been made in the capacity of VR to offer simulated environments which are quite acceptable from the sensory viewpoint).

The cultural potential of artifacts to shape and regulate human action – which, as we have already seen, is inherent in all kinds of artifacts – is explicitly displayed here: VR connects people in a very peculiar way and projects a different, alternative view of sociality. Its current applications, which are of considerable interest (for instance, in flight simulators for pilot training) only supply us with a pallid image of what, in the opinion of many, it will be able to provide in the near future.

This is not the place to examine the future prospects of VR from the standpoint of their technical feasibility. Our aim is not to assess the plausibility of the optimistic hypothesis that current technologies will shortly find satisfactory solutions to the problems, large and small, that still afflict VR applications – not least, that of motion sickness.

Here we wish to consider VR essentially as an environment of experience and of communication. It is true that VR may be a valid instrument for learning and manipulation, for teaching, scientific or operational aims – as, for example, in cases of surgical operations performed by remote control. However, its basic and most interesting characteristic is that it constructs artificial environments that actors can experience and in which they can communicate in new ways. As an environment that induces peculiar synthetic experiences and as an unusual communication medium, VR lends itself to several specific applications, which develop and at the same time exasperate the capacity of electronic communication to separate the interaction from interlocutors' physical co-presence. VR also embodies the idea – proper to the network paradigm – that communication consists in activating a potentially infinite and infinitely potent grid in which information and reality tend to become blurred.

The term 'virtual', which was previously used by computer scientists in expressions like 'virtual memory', began to gain ground in the late 1970s, when it was used to refer to highly interactive environments, such as 'virtual cockpits' for flight simulators or other special types of 'virtual workstations', also defined as 'virtual worlds' or 'virtual environments'. The feature common to all these environments was that the physical interface between computer and user was no longer limited to keyboard and screen. These relatively placid beginnings were the starting point for more adventurous developments.

The first to appear was 'artificial reality' (AR), which was developed in the mid-1970s by Myron Krueger (1990, 1991). He described AR as a technology aiming at allowing full-body participation in computer events in ways 'so compelling that they would be accepted as real experience' (Krueger, 1991, p.xiii), although in fact AR does not need to conform to the physical reality we experience in everyday life.

The 'cyber' approach is more recent. The term, coined by William Gibson, became famous thanks to the science fiction cult book *Neuromancer* (1984), although it had appeared two years earlier in another book by Gibson, *Burning Chrome*. But it was *Neuromancer* which really created a community of people united by an aspiration towards the development of a new kind of social interaction (Stone, 1991, 1992). For them, *cyberspace* is 'a parallel universe created and sustained by the world's computers and communication lines. A world in which the global traffic of knowledge, secrets, measurements, indicators, entertainments, and other-human agency takes on form' (Benedikt, 1991, p.1).

The expression 'virtual reality' (VR) was the last to appear. It was launched in 1989 by Jaron Lanier, first scientist and chairman of the board of VPL, a major manufacturer of VR in the United States. Industry's massive commitment to the development of VR hardware and software influenced its development to a considerable extent. However, although VR is the standard formula currently used to designate this area of applications, we should not forget that the range of experiences going under the name of VR is quite variegated and includes differentiated approaches (Biocca, 1992a,b).

AR highlights the advantages which artistic creative experience may draw from the construction of synthetic realities parallel to daily reality: 'The promise of AR is not to reproduce conventional reality or to act in the real world. It is precisely the opportunity to create synthetic realities, for which there are no real antecedents, that is exciting conceptually and ultimately important economically', says Krueger (1991, p. xiv). Instead, the cyber approach considers the development of electronic communication networks as an opportunity to go beyond our physical bodies and reach symbols and meanings directly. Although Krueger (1991) considers it as simply a special case of AR – a single AR which may be experienced simultaneously by thousands of people worldwide – cyberspace is much more than this for its adepts. It is 'a new and irresistible development in the elaboration of human culture and business under the sign of technology' (Benedikt, 1991, p.1), an environment of total communication accessible from any point on the earth's surface where there is a computer connected to the network.

In cyberspace, a social utopia, a philosophy and a religion all converge: a social utopia delineating the features of a free and disincarnate society; a philosophy going back to Plato, but also incorporating more recent ideas from Leibniz and Popper; and a religion which plainly identifies cyberspace with the City of God, where 'weightlessness, radiance, numerological complexity ... peace and harmony reign supreme' (Benedikt, 1991, p.15). Cyberspace stands as a mystic space of communication and belonging inside the citadel of the most advanced computer science. With the authority of scientific rationality and through the print of the MIT Press, it speaks of the 'erotic ontology of cyberspace' (Heim, 1991, p.59) and of a 'post-organic anthropology' (p.45) occupying a 'post-tribal cyberspace' frequented by 'shamanistic figures' (p.41). Reference to shamanism is quite pertinent, since the cyber adept aims to emulate the shaman's capacity to achieve visions which are not only true and illuminating but also useful and curative, by leaving his own body and travelling in spirit throughout the universe.

Instead, Lanier's perspective highlights the technological aspects of VR. He believes that, if technological research is as capable as it is ambitious, an immense market for VR applications will open up in the near future, making it 'the ultimate medium', the only communication environment capable of completing and complementing preceding ones. In spite of his extraordinary emphasis on the technological dimension, Lanier, curiously enough, complains that people are excessively dependent on technological models. He disapproves of the 'horrible substitution of information for human experience', imputing it to the fact that 'technology has been so overwhelmingly successful that it serves for most people as the most creative metaphor for what they are' (Lanier and Biocca, 1992, p.164). Readers may ask themselves whether a purely technological approach to VR – like the one adopted by Lanier – is enough to appreciate its underlying social and cultural processes. They may consider whether in the end emphasis on technology can be viewed as a cure for the 'horrible' disease of modelling human beings as technological objects, or whether it is perhaps a definite but surreptitious symptom of that sickness.

8.2   VIRTUAL WORLDS: MAPS OR TERRITORIES?

VR is featured by the media and understood by the public mostly as a techno-logical fact. People identify VR with a collection of machines: a computer capable of real-time animation, controlled by a set of wired gloves and a position tracker, and a head-mounted stereoscopic display for visual output. As a consequence, 'the focus of VR is technological, rather than experiential' (Steuer, 1992, p.73). A given system is classified as VR or non-VR according to whether it includes a minimum set of particular machines or not. The accent on technology, although it is understandable – because the quality of VR experience still leaves much to be desired, and expectations of substantial improvements rely heavily on tech-nological research – is disappointing to those who are interested in VR as a medium. They consider VR both as a way of using the computer and as a goal in resorting to it as a means of communication, rather than a physical device: early AR experiences, for example, exploited equipment quite unlike that which is currently used.

The problems posed by VR to researchers and professionals working in communications may be summed up in three questions: what form is this medium taking? how can we give it a form? and how will *it* form *us*? (Biocca, 1992a). These questions, which fit our approach to the interaction between actors and their everyday environments, consider VR as the link between technological devel-opments and new forms of sociality, between scientific programmes and religious aspirations, between conflicting desires to improve the existing world and to abandon it.

Before addressing these questions, let us examine two preliminary points. The first deals with the meaning of the word 'virtual'. As Woolley (1992) notes, it does not mean what many non-specialists think it means. The public at large appears to believe that 'virtual' reality is a special kind of reality, presuming that what is supplied by the helmet, gloves and general outfit is more or less equivalent to reality, i.e. 'virtually' the same thing (Shippey, 1993). In fact, 'virtual' means some-thing not truly real but only potentially so. People tend to confuse the dream – the mission, or rather the vision – of some research groups working on the development of new human–computer interfaces with something they imagine is already in being. This misunderstanding is by no means a small one, and may be followed in due course by disappointment. It is true that even expert researchers are sometimes over-indulgent in assessing the potential of their crea-tures, but we do know that they are quite aware of the fine line – and it may some-times be very fine – separating *virtual* from *real*. For example, when Shapiro and McDonald (1992) state that 'VR has the potential to involve users in sensory worlds that are indistinguishable or nearly indistinguishable from the real world' (p.94), we presume that they realize to what extent the word 'nearly' can be stretched and how great the step separating *potential* from *real* can sometimes prove.

The second point deals with the criticisms of current VR trends by those who are investing in a different kind of integration of the new technologies in daily life. Mark Weiser (1991), head of the Computer Science Laboratory of the Xerox Palo Alto Research Center, believes that the computer must be allowed 'to vanish into the background'. He believes that 'only when things disappear in this way are we free to use them without thinking and so to focus beyond them on new goals' (p.66). These goals usually do not consist (except for computer scientists) in knowing all about computers but in carrying out their daily activities as well as possible. But VR seems to be going in the opposite direction. It does not try to absorb the computer within everyday reality, but attempts 'to make a world inside the computer. Although it may have its purpose in allowing people to explore realms otherwise inaccessible – the inside of cells, the surface of distant planets, the information webs of data bases – VR is only a map, not a territory ... VR focuses an enormous apparatus on simulating the world rather than on invisibly enhancing the world that already exists' (*ibid.*).

We too believe that it is incorrect to substitute the map for the territory, since territory is only partially and diagrammatically represented on the map, for the reasons already discussed in previous chapters. But there are also other approaches (for instance, that of 'Mirror Worlds', MW) which share this centripetal view, according to which the world can and must be entirely absorbed into the computer. That is, when MW is up and running, people will be able to look inside their computers and see reality (Gelernter, 1991), a project that seems to be quite misguided. Both VR and MW are necessarily incomplete models of reality, they are not true alternative realities. They are within, not beyond, daily reality. Donald Norman (1990), a pioneer of studies on human–computer interaction, is quite critical of this kind of technocratic dreaming and thinks that, in the current developmental phase of human–computer interaction, less importance should be given to technologies and more to people, groups and social interactions.

8.3   EVIDENCE WITHIN ILLUSION: A PARADOX IN EXPERIENCE

The relationship with 'the already existing world' of everyday experience is critical for the quality of VR experience. What relationship exists between VR and a presumed reality, simply defined as such? In other words, exactly what type of reality is VR? What kind of experience can we encounter in VR? Considering the nature of these issues, Heim (1991) states that cyberspace is 'a metaphysical laboratory, a tool for examining our very sense of reality'. He asks: 'Does the meaning of "reality" – and the keen existential edge of experience – weaken as it stretches over many virtual worlds?' (pp.59–60). He believes that the answer to this question is *yes*, whereas we think that our sense of 'reality' may not be weakened but in fact sharpened, as VR forces us to clarify the epistemological positions we adopt, more or less consciously, in our everyday and scientific discourses.

On the one hand, the question of the relations between reality and VR is affected by the differences existing among the various current versions of VR. AR stands as a creative integration of ordinary reality. Cyberspace plots above daily experience, which it considers obsolete, like the physical bodies which inhabit it. Most technological research on VR apparently only deals with machines and programs, but perhaps it is technology itself which truly nurses the highest ambition of going beyond everyday experience and absorbing it without trace.

On the other hand, solutions to the question of the relations between reality and VR depend not only on how VR is viewed, but also on how 'reality' itself is conceived. We believe that, in general, VR makes it even more difficult for the 'ingenuous realists' to maintain their position. The very existence of VR tends to blur the line of demarcation between reality and VR only if this line is based on incorrect premises. The conflict between reality and VR is at its height for those who consider reality as defined by both the characteristics of being extra-mental ('objective', in the sense that it is 'outside' our heads) and of being based on sensory evidence (its vividness being the most plausible guarantee of its truth). This is precisely the case of those traditions of behavioural sciences which interpret theory as the result of the accumulation of 'objective' data gathered by means of experimental research which precedes theory itself and is independent of it. The 'ingenuous realism' of common sense, on which people rely in their daily lives, is also based on similar ideas. In the first chapters of this book we showed that this dualistic approach is not acceptable, because it fails to account for the very fact that in their ordinary lives actors face a reality in which 'inside' and 'outside' are closely intertwined.

From the point of view of the 'ingenuous realist', the two criteria which, taken together, define reality – extra-mentality and vivid sensory evidence – are mutually incompatible in VR environments. When related to the vividness of its sensory contents, VR experience appears persuasive and true; when regarded as a possibly reliable source of information about some 'external' reality corresponding to that sensory evidence, VR suddenly becomes illusory. This is disorientating for the 'ingenuous realist': in VR, Lanier says, 'the distinction between your own body and the rest of the world is slippery' (Lanier and Biocca, 1992, p.162).

There are of course other conceptions of reality which allow us to escape this baffling paradox; for example, Popper's idea of World 3 (Popper and Eccles, 1977) or Meinong's (1904) view of levels of reality organized in a series of layers, each supplied with its own conditions of existence, where no level is 'truer' than others – the level of existence on which Hamlet exists is only different from, but in no way less 'real' than, that inhabited by Queen Elizabeth during her historic life. Conceptions which consider social contexts as symbolic environments (Fulk *et al.* 1992; Spears and Lea, 1992) constantly constructed by actors are also part of this view. According to this constructivist perspective, possible confusion between various levels of reality is basically to be ascribed to actors' incompetence in dealing with the tools they use to construct and interpret their physical and social environments.

Just as we can easily imagine Don Quixote or Pinocchio as perfectly real personages and at the same time inhabitants of a peculiar fictitious world, so we can see no difficulty in considering the experiences offered by virtual environments as vivid and at the same time illusory knowledge. The reconciliation between the evidence of experience and the peculiar truth of the invention may occur inside a framework unlike that of 'ingenuous realism', which is still very widespread in many sectors of VR research. The conceptual category which allows us to overcome this discrepancy is 'fiction'.

If we think of reality and virtuality not in terms of 'true' (as given facts) as opposed to 'false' (as invented facts), but as fiction distributed over various levels of reality, we no longer have to deal with a baffling situation in which we cannot understand to what extent the VR experience is 'real'. We simply find ourselves faced with a series of invented worlds (Pavel, 1986), all socially constructed and all to be interpreted. Memory and imagination are kinds of reality – they belong to it, although they do not offer data but constructions. Romeo and Juliet are not blurred images of 'real' lovers in our daily lives, but living creatures in an invented world inhabited by certain personages, created by a certain playwright, read and acted by certain people.

If 'reality' stretches to embrace the worlds of invention, our idea of reality becomes wider and more clearly expressed. It contains the past, the present, and the future; and also the possible, the probable, and the improbable. Sancho Panza and Gulliver, Robin Hood, Othello and Jessica Rabbit, all fit into their respective niches.

8.4   CONSENSUS OVER HALLUCINATION: A PARADOX IN JUDGEMENT

The more the line of demarcation between reality and VR becomes blurred because VR succeeds in developing simulative environments acceptable from the viewpoint of sensory evidence (Schneiderman, 1992), the more difficult it becomes to express reality judgements which clearly distinguish between physical realities, computerized artifacts, computerized artifacts allowing physical reality to be manipulated, and so on (Ellis *et al.*, 1991b). At the level of judgement, the contrast between reality and VR corresponds to a similar opposition between correct reality judgements and hallucinations.

Psychological research on hallucination is based on the difference between self-generated states on the one hand, and external sources of information on the other. Discrimination between hallucinatory and non-hallucinatory states rests on the 'fundamental assumption that hallucinators mistake their own internal, mental, or private events for external, publicly observable events' (Bentall, 1990, p.88). Considered from the standpoint of hallucination, VR appears as an artificial environment which people enter with the aim of interacting with synthetically produced stimuli instead of 'normal' ones. The VR environment is sought as a source of experiences which are both sensorially vivid and illusory from the viewpoint of reality judgements. This type of situation seems to fall under the heading either of hallucination (Gibson considers VR as a 'consensual hallucination') or of non-hallucination (VR events are publicly observable, since virtual spaces may be shared by several actors).

The paradox is particularly baffling inside the perspective of 'ingenuous realism', which contrasts 'reality', supposedly located 'outside' people's minds, with 'illusion', which dwells 'inside' them. In fact, for those who follow this perspective, in VR all sorts of events have characteristics which make them similar to 'public events': they not only possess great sensory vividness but they also allow other actors, real or virtual, to take part, for example, in a game of tennis. Virtual environments are somehow always 'public' in the sense that they allow many different forms of telepresence, more or less efficiently simulated, from a mother grasping her faraway child's hand (Krueger, 1991) to the love-making between Casey and Linda in *Neuromancer*. However, VR is public in a sense that was foreseen neither by Bentall nor by common sense: both would prefer to consider it as a pseudo-hallucinatory context produced by technology – which can, incidentally, be accessed by other actors, human or simulated.

The underlying problem is still that of the first paradox of VR: what is the 'reality' status of VR experience? Who, or what, is the Linda that Casey meets? In Heim's (1991, p.65) view, Linda is to all intents and purposes a complete person with her 'sexual body' – 'the body and personality of Casey's beloved'. But this does not mean that she is not, or cannot be, a simulation produced by the master computer controller. Gibson's ambiguity refers to cyber's ultimate aim of transcending the corporeal world: 'The ultimate revenge of the information system comes when the system absorbs the very identity of the human personality, absorbing the opacity of their body, grinding the meat into information, and deriding erotic life by reducing it to a transparent play of puppets' (*ibid.*, p.66).

We can see that the confusion between private and public events arises not only from the sensory aspects of VR experience, but also from its very goal, which is that of exceeding and finally annulling ordinary life with its bodily bonds. We can now understand that the confusing effects of VR depend on two factors: (a) the sensory vividness of the presentation of the virtual world, and (b) the type of reality judgement which we are prepared to give on a metacognitive plane regarding VR experience. On this second level, two positions appear.

The first, which we will call 'moderately realist' but which many VR researchers prefer to call 'conventional', views the VR environment as an invented world, real in its own fashion but not to be confused with everyday reality. For supporters of this position, Linda is either Casey's girl in flesh and blood, or she is a simulation – the alternative being not irrelevant. They believe that, however efficient the simulation, we can and must always distinguish between the real and simulated Lindas, between the reality and the simulation, although we recognize that the simulation is in itself a sort of reality. For these 'moderate realists', VR is only a sensory hallucination, while reality judgements – although slightly disorientated – remain 'conventional': if she is a simulation, Linda is not 'really' real.

The second position, that of cyber movement, views the VR environment as a world which is not only not less real than the 'conventional' one; it is a world which perfects and goes beyond 'convention'. In this case, knowing whether the Linda who is embraced is 'conventionally' real or simulated is not of great importance, because the only difference – the physicality of her body – is a trap to be sprung. Technology allows the phase in which thought is imprisoned in organic, limited, obscure bodies to be overcome: 'the computer counterfeits the silent and private body from which mental life originated. The machinate mind disdainfully mocks the meat. Information digests even the secret recess of the caress' (Heim, 1991, p.66). If Linda is simulated, she is truer, in both body and personality, than if she were not: this is the core of the cyber credo.

For supporters of this metacognitive position, not only on the sensory plane but also and particularly on that of reality judgements, it is no longer possible to distinguish between 'conventional' and 'virtual' reality, favouring the former. The cyber orientation formulates 'a new definition of public and private: one in which warrantability is irrelevant, spectacle is plastic and negotiated, and desire no longer grounds itself in physicality' (Stone, 1991, p.106). This is the meaning of Gibson's stress on consensus: the cyber vision aims at building up a socially shared platform, making VR experience acceptable as a new and fitting communication environment, although it still leaves much to be desired as regards the quality of the sensory presentation.

On the level of reality judgements, the paradox of VR lies in the fact that, on the one hand, when consciously entering a VR environment we are aware that we are participating in a peculiar experience that is different from everyday reality. But on the other, and at the same time, at the moment of experiencing that environment, we place ourselves in the condition of not being able to discriminate appropriately between internal events and 'external' reality and undergo a deliberately induced hallucinatory type of experience. It must also be noted that VR is an environment in which it can be difficult *not* to hallucinate, at least slightly, as the 'contextual information' which is essential to formulate correct reality judgements (Bentall, 1990, p.88) is still rather poor. Especially for supporters of the cyber approach, VR is a deliberately disorientating metacognitive experience. This does not happen, or happens only to a lesser extent, when the context of its use, e.g. in flight simulators, supplies cues that anchor the VR experience to its proper social framework. In any case, the formulation of correct reality judgements can easily be regained through the awareness and memory of subjects that at a certain point they have entered a VR environment. Pragmatically (or do I mean paradoxically?) this corresponds to the possibility of testing the 'reality' of a VR environment simply by trying to get out of it.

It should also be noted that, on the sensory plane, current VR applications are far from being perfect (Earnshaw and Gigante, 1993; Wexelblat, 1993). The most disturbing limitations are the symptoms of motion sickness (Biocca, 1992c; Kennedy *et al.*, 1992; McCauley and Sharkey, 1992), the time lag in response, and the presentation of still over-simplified images, which are hardly acceptable when compared with those supplied by everyday situations or offered by media such as TV. At the present time, it is not difficult to distinguish between 'real' and simulated experiences, and the hallucinatory effect of VR remains modest. Although people working on the problem promise imminent wonders, we know that it is impossible for simulation to reach the level of complexity of the model (Rheingold, 1991).

What really happens is that, at least while their VR experience lasts, people attempt partial sabotage of the system presiding over the formation of reality judgements by doping their sensory channels with synthetically produced information that is anomalous with respect to the information to which they are bio-culturally adapted. People entering VR environments are not true hallucinators but only actors who – through the combined resources of both technology and fiction – aim at expanding everyday experience beyond the limits of dull routine where imagination is absent and sociality lacks socially recognized sense.

8.5   DUBIOUS IDENTITIES: A PARADOX IN COMMUNICATION

The visionary load of the cyber project cannot be understood unless two processes are borne in mind. The first is the spread of electronic communities, starting from the mid-1970s with the development of bulletin board systems (BBS), to the current emphasis on computer-mediated communication (CMC) as a stimulus to the development of democracy in organizations (Sproull and Kiesler, 1991; Rice, 1990; Mantovani, 1994b). The second process is people's growing isolation and the fragmentation of the social texture in the suburban technological 'parks' which came into being in the United States during the 1980s. *Neuromancer*, published in 1984, stands at the crossroads of these two processes: it provided an original myth to the growing aspiration towards alternative forms of community life and proclaimed VR an adventurous, new, free world. We would like to discuss here two points.

The first point is that VR is presented more as an environment of experience than simply as a vehicle for exchanging messages between people: 'Information is not transmitted from sender to receiver; rather, mediated environments are created and then experienced' (Steuer, 1992, p.77). However, the concept of communication in VR is not really so far from the traditional standpoint which views communication as a simple transfer of information: what Tatar *et al.* (1991) ironically call the 'parcel-post model' of communication. VR has in common with this model the idea that everything may be reduced to information transfer and information processing: in the new 'information society' we are faced simply with 'patterns of ideas, images, sounds, stories, data ... patterns of pure information' (Benedikt, 1991, p.4). By clinging to an obsolete idea of communication as mere transfer of information, VR jeopardizes its power to become a rich new communication environment.

The second point is the contrast between the physical success of telepresence, which in VR is provided quite efficiently, and interlocutors' true identities, which are becoming increasingly evasive and ephemeral. Steuer (1992) defines telepresence as 'the experience of presence in an environment by means of a communication medium' (p.75); its main feature is that this experience is mediated. In telepresence, the most relevant dimensions creating the sense of 'being there' are vividness and interactivity (Steuer, 1992, p.80). VR can offer vivid experiences of 'seductive tactile quality' (Heim, 1991, p.108) and interactivity as it is an 'intimate technology' (Krueger, 1991, p.250) which, for example, allows a mother to grasp the hand of her faraway child.

For most VR supporters, telepresence is a surrogate that is just as acceptable as physical presence. It can put people in contact either with other 'real' people or with synthetic realities. VR as a medium is a real or simulated environment in which people can experience telepresence. In both presence and telepresence, contact – understood in the sense of reciprocal sensory stimulation – is essential. Apart from its current imperfections, VR guarantees this contact. Instead – and it is here that its paradox as a medium arises – it can guarantee very little as regards identity. One disconcerting but widespread aspect of electronic communities is the assumption by conferencees of fictitious identities, mutually recognized as such, on the basis of temporary names ('handles'). Inside an electronic environment, one person may quite often construct several different identities and present her- or himself to others through several 'delegated puppet-agents' (Stone, 1991, p.105).

The metaphor of the computer as a theatre (Laurel, 1991) is literally applied here: in VR we encounter puppets which are manipulated for reasons unknown to some of its participants. Various factors, both individual and social, come into play when fictitious identities are developed. For example, the widespread use of attributing to oneself the identity of the other gender – female for men and male for women – may have a wide range of goals: knowing people of the opposite sex for possible dating, overcoming one's incapacity for establishing intimate, direct, interpersonal relationships, coming to know the feelings of people of the opposite sex, the pleasure of pretending to be someone else, and so on.

A clamorous and ethically worrying case, discussed by Stone (1991, pp.82–83; original source: Van Gelder, 1991), is that of Julie, a deserving member of an electronic community in the mid-1980s. Julie was a totally disabled elderly woman who wrote with a headstick and projected a personality full of interest for other people. Many women became her friends, began to confide in her and ask her for advice on private and life-influencing choices. Years later, thanks to the obstinacy of one of Julie's friends who wanted to meet her in person, although her handicap kept her in strict isolation, it was discovered that Julie was not a handicapped old lady at all, but a middle-aged male psychiatrist.

In the ensuing scandal, with protests from many women who had revealed their secrets to 'Julie', the psychiatrist explained the, to him, valid reasons why he had created this *persona* and made her work. This story aroused widespread interest in electronic communities during the late 1980s; other versions of it have been told, in which the elderly disabled Julie is transformed into Joan, a young neuropsychologist suffering from severe brain damage due to a car accident, while the puppet-handler, Arthur, remains a male psychiatrist in his early fifties.

The point is: did the fact that Julie did not exist change anything? For some women who later stated that they felt their privacy had been violated, whether Julie existed or not as a real person *was* important. For them, whether they physically knew her or not was not essential; what was essential to their relationship with her was the reality of her *identity*. This was the salient aspect. If Julie did not exist as a real person, the sense of their relationship with her did change: her words no longer had the same meaning if they came from another person. Conversely, other women were grateful to 'Julie' for her support and good advice. The 'real' source of her messages was not essential. These women separated the message from the identity of the interlocutor – typical of electronic communication. This is one of the changes in social conventions required by computer communications: 'many of the old assumptions about the nature of identity had quietly vanished under the new electronic dispensation' (Stone, 1991, p.83).

Every meeting in VR may present, sensorially amplified and complicated as regards reality judgements, the problems facing Julie's electronic friends when they learned that in one sense she did not exist and that in another she was a middle-aged male psychiatrist. Clearly, there are limitations to the simulation of an identity: an illiterate would find it practically impossible to pretend to be a literary critic and get away with it. However, the fact does remain that VR is a paradoxical communication environment in which the increasingly convincing presentation of a physically present interlocutor intertwines with an increasingly ephemeral 'presence'. Rheingold (1993) describes the vanishing of 'real' identities which takes place in CMC environments: 'Individuals find friends and groups find shared identities on-line, through the aggregated networks of relationships and commitments that make any community possible. But are relationships and commitments as we know them even possible in a place where identities are fluid? The physical world ... is a place where the identity and position of the people you communicate with are well known, fixed, and highly visual. In cyberspace, everybody is in the dark. We can only exchange words with each other – no glances or shrugs or ironic smiles. Even the nuances of voice and intonation are stripped away. On the top of technology-imposed constraints, we who populate cyberspace deliberately experiment with fracturing traditional notions of identity by living as multiple simultaneous personae in different virtual neighborhoods' (p.61).

We can now understand why Stone (1991) considers phone sex workers and VR engineers – people poles apart, one might think – as essentially working on the same problem, which is 'making humans visible in the virtual space'. 'The work of both is about representing the human body through limited communications channels, and both groups do this by coding cultural expectations as tokens of meaning' (p.102). Actually, visibility does not involve only the physical aspect but also the cultural acknowledgement of it: reciprocal recognition needs an at least partially shared symbolic order (Sahlins, 1985; Archer, 1988). Cultural context allows sensemaking processes to take place.

Something similar generally happens in CMC environments, as we saw in

Chapter 7, because in electronic environments social norms may function in stronger and less controlled ways than in interpersonal contexts in which inter-locutors are physically co-present. So we can expect VR, as a communication envi-ronment, to exacerbate the stereotyping of behaviours and fictitious identities, in order to assure the reciprocal intelligibility of actions and situations in a purposefully fictional environment.

8.6   WHAT KIND OF FICTION? INTERACTION VERSUS INTERPRETATION

But what is the benefit of thinking of VR as a fictional environment? And again, are we really sure that what inhabits virtual environments is the same type of fiction we find in traditional environments, for example, in literary invention? One advantage of considering VR in terms of fiction is that it frees us from the 'real'–'unreal' dichotomy, which is both conceptually misleading and culturally transient. Until a fictional environment is culturally well assimilated, its users feel the need to ask themselves how 'real' it is. For example, when the cinema first became popular, many novice spectators did not find it easy to accept the whistling locomotive hurtling towards them from the screen, and fled in terror.

So clarification of the fictional character of VR experiences may be useful, although the 'real'–'unreal' dichotomy, once it is acknowledged that VR belongs to fiction, turns out to be quite deceptive. This is because, in its own fashion, fiction is very 'real'. Children on the beach swapping sand-pies and pretending that they are real pies are not deceiving either themselves or anyone else. They are not stating that sand is edible and as sweet as sugar. They are simply devel-oping symbolic games in which their imagination fills up the void of 'real' ingre-dients, and they do not necessarily feel the need for them: a little boy who pretends he is fighting a crocodile does not necessarily wish to find himself facing a real one.

Daily experience is interwoven with the imaginary. Through fiction, imagina-tion feeds daily experience with repertoires of alternative possibilities, true parallel worlds that act as 'testing grounds' for mental exercises, as 'historical parks' for scenarios of projects and values, 'marginal models' directly flanking experienced reality and constantly influencing it (Pavel, 1986). The Natasha of *War and Peace*, who anagraphically does not exist, in the economy of the book and in readers' imaginations is more real than Napoleon. We know that Anna Karenina is a ficti-tious figure, but we consider that her story of love and despair enables us to under-stand real-life passions. Don Quixote is not the only one to confuse the real with the imaginary, if American *mafiosi* borrow their ways of dressing and of behaving in public from the cinema.

But while literary fiction stimulates the imagination, VR seems to aim at replacing it. A VR environment does not leave much space for the construction of a fantasy world: it offers an already constructed environment and invites its users to enter it and move about in it. This activity is completely unlike the one described above, in which children (and even adults) indulge when they make sand-pies or feed their baby dolls. 'I'm not a real doctor, but I play one in VR' (Shapiro and McDonald, 1992, p.94). Would we accept him as a doctor in real life? We are very far here from the game of the little girl who 'plays at being mummy' with her doll, dresses it up and hugs or scolds it. She is constructing a fantasy world; conversely, those who play at being doctors in VR are taking part in a training session in which they learn (or think they learn) some 'really' socially recognized professional ability.

Another advantage of resorting to the concept of fiction in VR is that it allows us to understand its specificity better by comparing it with other, pre-existing, well-known forms of fiction, such as literary fiction. The most important observation here is that literary fiction highlights interpretation: the pleasure of interpretation; the possibility of many competing interpretations (e.g. 'open works'). The worlds of invention in literature, similar to ambiguous everyday situations, acquire sense to the extent that they are interpreted in the light of our principles, aspirations and interests. VR, on the contrary, attributes little importance to interpretation. In VR environments, users are not asked to construct, equip with meaning and interpret situations, as in literary fiction; they are stimulated to interact with the environment, to do something, to touch something (Stone, 1992). VR offers environments which are as complete as possible, within the limits fixed by available technological resources – environments without voids, full of 'true' facts. In VR, grasping a tennis racket is as real as ducking to avoid a pterodactyl swooping down. We know they are not 'really real', but this does not stop VR environments from presenting themselves as sensorially well-documented evidence to which we must respond by doing something in a physical sense. In VR, imagination and interpretation play a purely marginal role: they dwell in nooks and crannies, in the blank spaces left where the synthetic environment failed.

A third advantage in resorting to the idea of fiction is that it reveals interesting differences between VR environments and traditional literary fiction with respect to such features as distance, moral complexity, and secrecy.

Let us consider *distance*. A written text presents us with a very incomplete story: how does Dr Zhivago's Lara smile? how does she move? The text is generally quite indeterminate, because the book is a 'lazy machine' that is always requesting readers to do part of its job (Eco, 1994). In a book, what is lacking physically is supplied by the reader in a complex game of identifications, categorizations and projections. The allure of the written text lies mainly in its incompleteness, in the gap it maintains between what is printed and what can only be imagined. In contrast, VR environments are neither distant nor incomplete, at least not purposefully. They aim at overcoming physical distance; about symbolic distance they care little. Lanier even believes that, since signs and symbols only exist to overcome our incapacity to manage physical realities, in VR 'communicating without codes' is possible because 'VR – although it's a low quality reality – is the only other one that's truly objective', apart from physical reality (Lanier and Biocca, 1992, p.160).

Lanier clearly adopts here a viewpoint inspired by the above-mentioned 'ingenuous realism', which makes it deceptively easy for him to distinguish what is 'objective' (physical nature) from what is not (everything else, except VR). Technologists' visions sometimes fall short of their achievements. This is one of the reasons why it is not advisable for the new media to be left only in the hands of technologists – actually, there is no such thing as a 'pure' technologist. People committed to technology are, like all other socio-professional clusters, social activists equipped with a definite 'moral philosophy' (Dunlop and Kling, 1991).

As regards *moral complexity*, we see that it is an essential part of any written narrative text (Eco, 1979), which consists in the construction of a character, a situation or a social context in terms of symbols, principles, and values. Literary works are compelling or even thrilling tales often based on thin premises. They captivate their readers (Walton, 1984) by means of processes of identification with their characters, persuading them to attribute some kind of meaning to the narration, to make it their own (Bruner, 1986, 1990). Instead, rather than presenting symbolic contexts, VR aims at effectively simulating the external appearances of environments. Rather than stimulating in them the need to make sense of ambiguous – and morally interesting – situations, it offers its inhabitants an opportunity for the physical manipulation and exploration of synthetic universes. More than attributing meanings, it is interested in interactivity as manipulation.

As for *secrecy*, we know only too well that it is essential in written texts which are really important, for example, sacred writings. The message that a holy book conveys must be decoded and accessed privately, by people able and worthy of the revelations it contains. In sacred writings, secrecy is an essential part of the message (Kerkmode, 1979). The text reveals and conceals at one and the same time. It speaks in parabolas, so that the uninitiated, although capable of hearing, are excluded from understanding of the deep, hidden message that is reserved for the elect. In VR environments, there is no discipline of secrecy, there are no degrees of election.

What the VR environment offers is manifest: if a distinction can arise, it is not between adepts and non-adepts, but between those who are expert and those who are inexpert, between those who are capable and those who are incapable of fully exploiting the potentialities of the new media. In place of the traditional hierarchy discriminating between readers, composed of graded steps towards truth, VR provides a hierarchy based on the possibility of access and of competences in controlling the new technologies.

The peculiar power of written texts – their imaginative and dramaturgical wealth – is confirmed by an interesting phenomenon: the spread in working environments and as a social game of a class of text-based virtual world systems, called multi-user domains (MUDs). This deliberately unfortunate acronym, officially deriving from a 1979 computer game called Multi-User Dimension, also stands for Multi-User Dungeons, a game of adventure and fantasy which has been very popular on American campuses for years. MUDs use text instead of physical appearance; participants have to face the fact that their experience is reduced to pure text by the medium, which obliterates all forms of social interaction based on physical proximity. The result is increasing interpretative ability: 'MUD users are able to read between the lines of text that make up their virtual world' (Reid,

1995, p.173). In this, MUDs are singular environments because they develop interest in interpreting texts rather than in manipulating objects, although the tendency to manipulation is still very strong (e.g. the need to express strong mutual feelings can induce participants to use textual information in ways that make possible the construction of physical structures simulating, for instance, a home to share, or pets and children with whom to interact).

MUDs successfully seek to appropriate for themselves some of the advantages of written texts, such as their moral complexity. They are places where actors can play freely with social roles, identities and self-disclosure. Gender issues are quite central in these text-based virtual environments. Seemingly, this also has grammatical reasons: 'All MUDs allow, some insist, that players set their "gender flag", which controls which set of pronouns are used by the MUD program in referring to the player. Most MUDs allow only three choices – male, female, and neuter – which decide between the families of pronouns containing him, her, or it. A few MUDs demand the players to select either male or female as their gender and do not allow those with an unset gender flag to enter the MUD. Other MUDs allow a great many genders – male, female, plural, neutral, and hermaphrodite being among the possibilities' (*ibid.*, p.179).

Obviously, emphasis on gender is produced by the fact that VR environments are spaces in which people can and want to experiment with gender-specific social roles, i.e. males play female characters, or vice versa. Discussion about the ethical problems involved are normal in MUDs, but here – unlike the above-mentioned case of Julie's friends who were confused and upset when they learned that she was and at the same time was not a totally disabled elderly woman – participants know the basic rule of CMC very well: 'whether an individual user enjoys the situations that come of this potential or is resentful and wary of them, all are aware that exploitation of it is a part of the MUD environment' (*ibid.* p.181).

MUDs, born as interactive role-playing games, recently developed into rather effective tools for communication within work groups. They are virtual 'places' on the network where members of a team, wearing their particular 'characters', can meet in a shared context. This is usually organized into 'rooms', spatial metaphors being crucial in building MUDs. In such places, participants can talk to each other, addressing either people present in the room or, using a particular 'paging' command, to those located outside it. Some of the features of normal conversation can be reproduced in MUDs – 'whisper' command, for example, makes it possible to talk privately with a single interlocutor. MUDs try to build a shared space not only in its physical sense, but also in its social sense: they have a history mechanism, they are exclusive to team members, and they allow situation problem solving.

8.7   POSSIBLE SELVES AND THE SOCIAL NEGOTIATION OF IDENTITY

We have seen that VR is an environment both exciting and confusing, inhabited by paradoxes regarding experience, reality judgements and communication. We have also seen that VR, as a fictional world, calls for manipulation rather than interpretation. Like the experience of television but to a far greater extent, VR tends to separate vividness from connotation, presence from identity, meaning from social and cultural context (Bryant and Zillman, 1991; Kubey and Csikszentmihalyi, 1990). What effects can we expect if VR spreads as television did? When answering this question, we must bear in mind that, 40 years ago, few experts anticipated the massive impact that TV would have – and has had – on people's lives.

We believe that the new media not only influence people's choices, thanks to their capacity for persuasive images, but that they also contribute greatly to moulding their identities (McQuail, 1994). Recourse to the concept of possible selves may be useful in understanding this aspect of the media. 'Possible selves give specific cognitive form to our desires for mastery, power, or affiliation, and to our diffuse fears of failure and incompetence' (Markus and Nurius, 1986, p.960). Possible selves are repertoires of metaprojects and most persistent motivations. Our individual freedom to plan or imagine the future is limited not only by given symbolic order, but also and particularly by the images we receive through the media.

In the postmodern world, the media make actual that symbolic order (Giddens, 1991) which would otherwise remain remote and empty. It is through them that we perceive our social context. The less mediation is detected, because the medium is taken as equivalent to the 'natural' one, the more mediated experience is surreptitiously presented and tacitly accepted as direct experience of 'natural' environments, without people becoming aware of the ways in which mediation works. From this viewpoint, the influence of the new media in guiding the construction of possible selves must not be underrated. They perform an essential function in cultural socialization: 'Possible selves have the potential to reveal the inventive and constructive nature of the self but they also reflect the extent to which the self is socially determined and constrained' (Markus and Nurius, 1986, p.954).

A further reflection shows that the encounter between media and possible selves occurs inside a very 'private' space, mainly kept hidden from social negotiation. We usually think of the self and of people's long-term projects as quite stable constructions (Abrams, 1992; Deaux, 1992). Possible selves lead us into a private place in which the self is not yet completely formed and plans still have to be made. Other people react to publicly accessible manifestations of our identity, but they are generally unaware of our ideas regarding our own potentials, positive or negative, expressed as alternative personal identities. As a consequence, possible selves are poorly controlled by the social context surrounding us in everyday life: 'Individuals have ideas about themselves that are not well anchored in social reality ... as representations of the self in future states, possible selves are views of the self that have not been verified or confirmed by social experience' (ibid., p.955). Individuals alone know and decide about their possible selves, in relative secrecy and in isolation from others, although always under the influence of the normative social system in which they participate.

Can we thus say that people exposed to the influence of manipulative environments such as TV and VR – capable as they are of convincingly mimicking important aspects of 'reality' without the person immersed in them being in full control of the experience negotiated by the medium – may not be fully aware of the normative influences to which they are subjected through that medium? We believe the answer must be a definite *yes*.

At this point, evaluation of the impact of VR requires closer analysis in terms of social psychology, education and personality development. Until now we have dealt with the impact of media on people's behaviour; perhaps now we must begin to worry about their impact on people's imaginations. The media are 'dynamic sites of struggle over representation, and complex spaces in which subjectivities are constructed and identities are contested' (Spitulnik, 1993, p.296). We used to consider media merely as channels that convey information, but we may now begin to think of them as dispensing norms as well, which can direct people's choices. Since the interactivity of VR cannot be compared with the situation of exposure to TV, we may expect VR to have even greater powers of suggestion than those of TV. These powers are increasingly less counterbalanced by people's direct experience, especially young people, and by 'conventional' social contexts. The latter are beginning to vanish from processes of identity construction (Taylor, 1989, 1991), particularly in the young, because of the withdrawal from their lives of the influence of those traditional ethical communities – like family, school, church – which until very recently exercised not only an informational but also a strong normative influence.

To the extent that the supply of information and images will increase without users' corresponding capacity for integration and synthesis, the development of new media may provide opportunities for the formation of increasingly numerous possible selves, always more divergent and always more fragmented. Information, if it is to be used, has to be organized into meaningful chunks taking the form of narrative structures, of stories to be told and shared among the members of a group (Bruner, 1990). To the Babel of incoming information, VR may only add sensory vividness and the possibility of manipulating images and of acting inside the medium. New questions arise: what difference is there between watching an erotic film and indulging in sexual activities in VR, or between watching a scene of violence on TV and killing another person in VR? VR has thrown down the gauntlet of challenge, and we are now being required to demonstrate our capacity to reconstruct situations inside a symbolic order which confers plausible shared meanings upon them. This challenge may be overcome, but we must be conscious of its nature and stature, especially as regards the capacity of these new environments to manipulate our imagination processes. Imagination is at the very basis of our moral reasoning (Johnson, 1993), as we rely on a set of metaphors (implicit or explicit) to structure the problem space so that our responses to problems are highly dependent on the choice of the metaphor we use to frame a given situation.

We believe that the hallucinogenic potential of VR must not be underestimated. In human cognitive functioning, there is a very strong tendency to accept incoming information, particularly sensory, as true: in mental systems 'acceptance occurs prior to or more easily than rejection', because 'unacceptance is a more difficult operation than acceptance' (Gilbert, 1991, p.111). This may be explained in evolutionary terms: we act in a world in which it is important to respond promptly to situations, while accuracy does usually not have top priority. The result is that human cognitive systems have developed adaptively 'the tendency to treat all representations as if they were true' (ibid., p.116), except when there is proof to the contrary. But it is not always possible to find proof to the contrary, especially in environments in which information is multiple and discrepant and there is little time to evaluate it properly, as with modern media. TV, and possibly to an even greater extent VR, can submerge people under a flood of information which exceeds their attentional resources and capability for filtering.

The enormous human hereditary potential for acquiescence in given data – and sensory data in particular – is probably underestimated, both in studies on VR and in those on the effects of exposure to TV (Bryant and Zillman, 1991). The distinction between reality and illusion is quite blurred in VR environments; it is less so in TV (Fitch et al., 1993). Is this not precisely one reason for their appeal? The increasing spread of the new media accelerates the dissolution of traditional normative contexts, whose withdrawal risks emptying those media of points of reference and meaning. TV shows already speak almost exclusively of what happens in the world of the small screen. Will something of the kind happen in the VR of tomorrow?

This vicious circle could be interrupted by the development of something currently lacking, a culture capable of conferring proper meanings on the experiences which are made in and through the new media. The latter must not be exorcised, but interpreted through conceptual categories which may prove appropriate, more mature, than the ingenuous social utopias expressed until now in VR environments. The ideas of progress, efficiency and democracy which still accompany the spread of informational artifacts may become not vain embellishments, but the very substance of the projects implicit in the new technologies (Dunlop and Kling, 1991). We can criticize and develop these projects, purifying them of their present shortcomings – first and foremost the unhappy cyber metaphor of the human–computer marriage. The cyber idea that 'our affair with information machines announces a symbiotic relationship and ultimately a mental marriage to technology' (Heim, 1991, p.67) is dangerous for the development of our identity and misleading for an understanding of VR.

We have to acknowledge that from its very beginnings, VR looked for new, intimate social bonds: 'Virtual communities might be real communities, they might be pseudocommunities, or they might be something entirely new in the realm of social contracts, but I believe they are in part a response to the hunger for community that has followed the disintegration of traditional communities around the world' (Rheingold, 1993, p.62). The need for belonging, which animates most CMC environments, is given a peculiar response by the fact that computers are 'linking' machines, capable of overcoming physical barriers but less able to build shared meanings. The problem in CMC is that, 'once we can surmount time and space and "be" anywhere, we must choose a "where" at which to be, and the computer's functionality lies in its power to make us organize our desires about the spaces we visit and stay in' (Jones, 1995, p.32).

This *where* is more a cultural than a physical dimension, as we saw in the chapters on actors' situated interests and goals. People who seek community and self-fulfilment by using CMC may be deluded: no one medium, no one technology *per se* can provide a novel, stimulating or at least decent, social context to replace the old one which is so quickly vanishing.

CHAPTER NINE

Conclusions

Electronic Communication, Social Context and Virtual Identities

Let us conclude this work with a few brief observations, both retrospective and prospective. If we turn back to examine the path we have been following, we can see both the direction and the internal consistency of our discourse. We started from the new cognitive paradigm of situated action, passed through the presentation of a conceptual model of the social context and of artifacts as embodied projects, and came to consider the new environments of cooperation and communication produced by information technology. Through a series of explicit internal links, we have gradually applied and revealed our reference grid, inspired by situated action, which we used to capture the specific features of the new environments.

In Part 1 of this book we claimed that everyday situations are inherently ambiguous, not because actors lack information but because that same information, if it is to have any meaning, must be interpreted. Actors' interpretations of situations depend on the changing system of their interests which compete for priority in access to available resources. If we examine the relation linking actors' interests to environmental affordances, we see why situations are structured in various ways not only by various actors, but even by one actor at various moments in time.

Understanding that actors inevitably develop mutually irreconcilable points of view about the world is essential for a realistic approach to the new communication environments, in which cooperation cannot be separated from conflict and negotiation among different points of view. According to our perspective, the emergence of conflict and negotiation is explained not only in fact but also in principle, thus allowing us to overcome one of the limitations noted in many studies on computer-mediated communication (Castelfranchi, 1992) and computer-mediated cooperative work – that of considering work as mere cooperation.

The cognitive and social perspective presented in Part 1 of this book is based on the innovative but already classical theories of Agre, Beach, Clancey, Greeno, Lazarus, March and Suchman. It aims at overcoming the dichotomy between 'inside' and 'outside', between actors and environments, which still afflicts wide areas of psychological research. Emphasis is now placed on the ongoing interaction taking place between actors and environments.

Context is the concept which becomes pivotal in our approach. What is context? We can easily acknowledge that context cannot be considered as a given fact, but as a multiple space for interaction and interpretation: 'like "rationality", the continuity of activity over contexts and occasions is located partly in person-acting, partly in contexts, but most strongly in their relations' (Lave, 1988, p.20). How does this common and constantly changing area, in which actors and situations meet by means of interpretational processes, come into being? And how does it function?

In Part 2 of this book we proposed a three-level model of social context as a framework responding to these questions. We did this by following two inverse but complementary pathways. The first is top-down, and goes from symbolic order to everyday practices: actors resort to systems of principles and cultural models in order to give significance to their actions and to attribute sense to others' behaviour. It is the presence of symbolic order as a pre-existing repertory of shared meanings that allows actors to communicate and to take on social roles as webs of mutually recognized and binding commitments. Ritual, counterposing what is done in everyday reality with what should be done if things functioned properly in an ideal world, shows that symbolic order is both impossible to achieve and independent of everyday experience. Ritual is not just a copy or an idealized version of real life: it criticizes it, starting from a set of meanings which are transmitted by tradition, accepted in the present by actors, and constantly reproduced and altered by situated action.

The second pathway is bottom-up, from practices to meanings, from artifacts to the goals actors try to achieve. Technological artifacts, such as the instruments produced by the new technologies, are not just neutral passive tools. They are living things, they embody values and social goals – often implicit but not any less well defined – which we must learn to recognize. Not only technological artifacts but also what we (incorrectly) call 'cultural' artifacts (incorrectly, since technological instruments are themselves cultural products) are vehicles conveying individual and social projects.

What are these projects? Who formulates them? What aims do they have? The twin issues of power and technology are closely connected, although they are rarely faced directly and explicitly. The phrase 'the politics of artifacts' is not an inane label; it stresses the often hidden potential of technology to modify and finally control people's conduct. If we turn to the environments of electronic cooperation and communication, it is easy to see how much they influence our everyday activities.

Actors' responsibilities towards the technologies which permeate our living and working environments – which is the topic closing Part 2 of this work – still require not only greater accessibility and transparency of artifacts, as already noted in our previous work on the quality of human–computer interaction (Mantovani, 1994a). They also require a capacity for cultural processing on the actors, appropriate to the new situations which are arising as the new communication environments advance.

Part 3 of this book is devoted to a critical analysis of these environments: the 'network paradigm' and its limitations, the arguable myth of electronic democracy, and the supposed role of 'shared spaces' and 'community handbooks' as catalysts for the development of 'electronic altruism'. Central to Part 3 is a discussion of the relationship between the social dimension of computer-mediated communication and the idea of context as symbolic order. This is one of the points we consider most important, because it shows how our model of context can contribute to an understanding of the nature of these new environments.

To conclude, two points extend beyond the confines of the discussion so far.

The first deals with the theme of identity, which underlies the whole of our discourse and emerges from time to time: for example, when discussing image theory in intuitive decisions, social identity theory, or interpretation of situations from the viewpoint of a person sitting in front of a computer keyboard and wishing to communicate with other people. But it is in the field of virtual reality that the question of identity really comes to the fore. The cyber perspective, as we have seen, seeks from the new communication technologies access to new forms of sociality, freed from the body yet still sensorily vivid. It also seeks different ways of presenting identity and developing relationships, ways which combine efficient simulation of the presence of other people with deliberate uncertainty concerning the real identities of interlocutors.

The paradoxes of telepresence seem to predict virtual worlds populated by equally virtual identities, vivid but illusory images. Are the new media pushing us, slowly but inexorably, into a universe inhabited by actors wearing multiple personalities, to be put on and taken off at will, like the fictitious identities which the participants of many electronic communities like to assume? Although human beings have always used fiction to enrich their daily experiences, we must ask ourselves now whether the fiction of the new media is similar, in its nature and effects, to the forms we already know, to which we are accustomed, and from which we extract the best – or at least, so we think when we read a moving novel or watch a good western.

What is the social and cultural context in which we are developing electronic communications environments, which may fire our imaginations but about which we really know too little? Only now are we beginning to ask ourselves the right kinds of questions – for example, about TV as an environment in which people can remain immersed for long periods of time, in some cases in a sort of addiction which mimics the condition of altered states of consciousness.

Such questions on the one hand reveal the need for deep research into these relatively unexplored areas; on the other, they already contain within themselves some of the answers which we are today able to give. We do not ask ourselves *where* the new technologies are leading us, but *what* we are doing with them, towards what goals we are guiding them. Technologies lie within the social context, not vice versa. They are parts of actors' projects, not realities in themselves. But for this very reason they require a capacity for cultural interpretation that meets the challenge of the situation (Gergen, 1991).

The second point deals with the effect of contexts on theories. We have seen how the development of electronic communication environments defies some of the tenets of the traditional theory which considers communication simply as the transfer of information. The traditional theory, of cybernetic origin, is increasingly being rejected because it cannot explain important aspects of the new environments. These are captured by more recent and more social concepts which consider communication as a process in which shared meanings are negotiated. This paradigm shift is interesting both for its intrinsic relevance to communication research and because it confirms some of the epistemological principles stated by McGuire's 'contextualism', which views contexts not as passive objects to which theory-formulated predictions can be applied, but as environments eliciting the potentials and limitations of scientific theories. We too believe that no theory can exhaust the complexity of real-life, everyday situations.

So we agree with McGuire's (1983) idea that 'theory – like knowledge on any other level – is an oversimplified and distorted representation of any situation. It can be a brilliant cost-effective representation in certain contexts and dangerously misleading in others. Because all hypotheses are true, all are false. A hypothesis or its contradiction is each adequately true in a few appropriate contexts and each is dangerously false in many others' (p.7). Verification of a theory, understood as progressive clarification of the hypotheses it contains and generates, occurs by means of its contextualization, i.e. precise reference to various situations, with respect to which it reveals various degrees of validity. The meaning of a theory is discovered 'by revealing its hidden assumptions and so specifying to which contexts its misrepresentations are tolerable and in which seriously misleading' (*ibid.*, p.8).

As attentive readers will no doubt have noted, we have now come back to our point of departure, that is, the complexity of situations. Have we been forced to drop back, like unlucky players in a game of 'Snakes and Ladders'? Or, to change metaphors, have we galloped ahead, passed the winning post and won the prize, to be jealously hoarded and used to bet successfully on the next race? This, after all, is what we really want.

What is the next game we will play, what is the next race we will run in the coming years, in our living and working environments, when the new technologies have improved beyond all expectations? Whatever happens, the game – once again – will have to be played on the levels of both symbolic order and artifacts, on the planes of theoretical conceptions and empirical research. And we will need – then as now – strong points of reference in order to understand and direct our use of the new environments of communication.

References

ABRAMS, D., 1992, Processes of social identification, in Breakwell, G.M. (Ed.) *Social Psychology of Identity and the Self Concept*, pp.57–100, London: Surrey University Press.

ABRAMS, D. and HOGG, M.A., 1990, An introduction to the social identity approach, in Abrams D. and Hogg M.A. (Eds) *Social Identity Theory*, pp.1–9, Hemel Hempstead: Harvester Wheatsheaf.

ACKERMANN, D. and TAUBER, M.J. (Eds), 1990, *Mental Models and Human–Computer Interaction*, Vol. 1, Amsterdam: North-Holland.

ADRIANSON, L. and HJELQUIST, E., 1991, Group processes in face-to-face and computer-mediated communication, *Behaviour and Information Technology*, **4**, 281–296.

AGRE, P.E., 1990, Review of 'Plans and situated actions' by Lucy Suchman, *Artificial Intelligence*, **43**, 369, 384. Reprinted in Clancey, W.J., Smoliar, S.W. and Stefik, M.J. (Eds), 1994, *Contemplating Minds*, pp.223–241, Cambridge, MA: MIT Press.

AGRE, P.E., 1993, The symbolic worldview: Reply to Vera and Simon, *Cognitive Science*, **17**, 61–69.

AGRE, P.E. and CHAPMAN, D., 1987, Pengi: An implementation of a theory of activity, *Proc. Sixth National Conference on Artificial Intelligence*, pp.268–272, Menlo Park, CA: American Association for Artificial Intelligence.

AHRNE, G., 1990, *Agency and Organization*: *Towards an Organizational Theory of Society*, London: Sage.

ALEXANDER, J., 1988, *Action and its Environment*, New York: Columbia University Press.

ALTMAN, I., 1979, Privacy as an interpersonal boundary process, in Cranach M., Foppa, K., Lepenies, W. and Ploog D. (Eds) *Human Ethology*, pp.95–132, Cambridge: Cambridge University Press.

ANDERSON, R.J., HEATH, C.C., LUFF, P. and MORAN, T.P., 1993, The social and the cognitive in human–computer interaction, *International Journal of Man–Machine Studies*, **38**, 999–1016.

ANDERSON, R. and SHARROCK, W., 1993, Can organizations afford knowledge?, *Computer-Supported Cooperative Work*, **1**, 143–161.

ARCHER, M.S., 1988, *Culture and Agency*, Cambridge: Cambridge University Press.

ARENSBURGER, A. and ROSENFELD, A., 1995, To take arms against a sea of e-mail, *Communications of the ACM*, **38** (3), 108–109.

AUSTEN, J., 1986, *Sense and Sensibility*, London: Penguin (1st edn 1811).

AUSTIN, L.C., LIKER, J.K. and McLEOD, P.L., 1993, Who controls the technology in group support systems? Determinants and consequences, *Human–Computer Interaction*, **8**, 217–236.

BAGNARA, S., ZUCCHERMAGLIO, G. and STUCKY, S. (Eds), 1994, *Organizational Learning and Technological Change*, Berlin: Springer.

BANNON, L., 1991, From human factors to human actors: The role of psychology in HCI studies in system design, in Greenbaum, J. and Kyng, M. (Eds) *Design at work*: *Cooperative Design of Computer Systems*, pp.25–44, Hillsdale, NJ: Erlbaum.

BANNON, L. and SCHMIDT, H., 1991, CSCW: Four characters in search of a context, in Bowers, J.M. and Benford, S.D. (Eds) *Studies in Computer-supported Cooperative Work*, pp.3–16, Amsterdam: North-Holland.

BARA, B., 1995, *Cognitive Science and Developmental Approach to the Simulation of the Mind*, Hillsdale, NJ: Erlbaum.

BARGH, J.A., 1990, Auto-motives: Preconscious determinants of social interaction, in Higgins, E.T. and Sorrentino, R.M. (Eds) *Handbook of Motivation and Cognition*, Vol. 2, pp.93–130, New York: Guilford.

BARON, R.M. and BOUDREAU, L.A., 1987, An ecological perspective on integrating personality and social psychology, *Journal of Personality and Social Psychology*, **53**, 1222–1228.

BEACH, L.R., 1990, *Image Theory*: *Decision Making in Personal and Organizational Contexts*, Chichester: Wiley.

BEACH, L.R., 1993, Broadening the definition of decision making: The role of prechoice screening of options, *Psychological Science*, **4** (4), 215–220.

BENEDIKT, M., 1991, Introduction, in Benedikt, M. (Ed.) *Cyberspace*: *First Steps*, pp.1–25, Cambridge, MA: MIT Press.

BENIGER, J.R., 1990, Conceptualizing information technology as organization, and vice versa, in Fulk, J. and Steinfield, C. (Eds) *Organizations and Communication Technology*, pp.29–45, Newbury Park, CA: Sage.

BENTALL, R.P., 1990, The illusion of reality: A review and integration of psychological research on hallucinations, *Psychological Review*, **107** (1), 82–95.

BERREMAN, G., 1966, Anemic and emetic analysis in social anthropology, *American Anthropologist*, **68** (1), 346–354.

BIKSON, T.K., 1987, Understanding the implementation of office technology, in Kraut, R. (Ed) *Technology and the Transformation of White Collar Work*, Hillsdale, NJ: Erlbaum.

BIKSON, T.K. and EVELAND, J.D., 1986, *New Office Technology*: *Planning for People*, New York: Pergamon Press.

BIKSON, T.K. and EVELAND, J.D., 1990, The interplay of work group structures and computer support, in Galegher, J., Kraut, R.E. and Egido, C. (Eds) *Intellectual Teamwork*: *Social and Technological Foundations of Cooperative Work*, Hillsdale, NJ: Erlbaum.

BIKSON, T.K., EVELAND, J.D. and GUTEK, B.A., 1989, Flexible interactive technologies for multi-person tasks: Current problems and future prospects, in Olson, M.H. (Ed.) *Technological Support for Work Group Collaboration*, Hillsdale, NJ: Erlbaum.

BIOCCA, F., 1992a, Communication within virtual reality: Creating a space for research, *Journal of Communication*, **42** (4), 5–22.

BIOCCA, F., 1992b, Virtual reality technology, *Journal of Communication*, **42** (4), 23–72.

BIOCCA, F., 1992c, Will simulation sickness slow down the diffusion of virtual environment technology?, *Presence: Teleoperators and Virtual Environments*, **1** (3), 334–343.

BØDKER, S., GREENBAUM, J. and KYNG, M., 1991, Setting the stage for design as action, in Greenbaum, J. and Kyng, M. (Eds) *Design at Work: Cooperative Design of Computer Systems*, pp.139–154, Hillsdale, NJ: Erlbaum.

BOWERS, J., 1992, The politics of formalism, in Lea, L. (Ed.) *Contexts of Computer-mediated Communication*, pp.232–261, Hemel Hempstead: Harvester Wheatsheaf.

BOWERS, J. and BENFORD, S. (Eds), 1991, *Studies in Computer-supported Cooperative Work*, Amsterdam: Elsevier.

BRENNAN, S.E., 1991, Conversation with and through computers, *User Modeling and User-adapted Interaction*, **1** (1), 67–86.

BRONFENBRENNER, U., 1979, *The Ecology of Human Development*, Cambridge, MA: Harvard University Press.

BROWN, J.S., 1986, From cognitive to social ergonomics and beyond, in Norman, D.A. and Draper, S.W. (Eds) *User Centered Systems Design*, pp.457–486, Hillsdale, NJ: Erlbaum.

BROWN, J.S. and DUGUID, P., 1994, Borderline issues: Social and material aspects of design, *Human–Computer Interaction*, **9**, 3–36.

BRUNER, J., 1986, *Actual Minds, Possible Worlds*, Cambridge, MA: Harvard University Press.

BRUNER, J., 1990, *Acts of Meaning*, Cambridge, MA: Harvard University Press.

BRUNER, J, 1993, Do we 'acquire' culture or vice versa?, *Behavioral and Brain Sciences*, **16**, 515–516.

BRYANT, J. and ZILLMAN, D. (Eds), 1991, *Responding to the Screen*, Hillsdale, NJ: Erlbaum.

BUSS, D.M., 1991, Evolutionary personality psychology, *Annual Review of Psychology*, **42**, 459–491.

CARNEVALE, P.J. and PRUITT, D.G., 1992, Negotiation and mediation, *Annual Review of Psychology*, **42**, 531–582.

CARROLL, J.M. and CAMPBELL, R.L., 1989, Artifacts as psychological theories: The case of human–computer interaction, *Behaviour and Information Technology*, **8** (4), 247–256.

CARROLL, J.M., MACK, R.L. and KELLOGG, V.A., 1988, Interface metaphors and user interface design, in Helander, M. (Ed.) *Handbook of Human–Computer Interaction*, pp.67–85, Amsterdam: North-Holland.

CASTELFRANCHI, C., 1992, No more cooperation, please! In search of the social structure of verbal interaction, in Ortony, A., Slack, J. and Stock, O. (Eds) *Communication from an Artificial Intelligence Perspective*, pp.205–228, Berlin: Springer.

CASWELL, S.A., 1988, *E-mail*, Boston: Artech House.

CIBORRA, C. and Lanzara, G.F., 1990, Designing dynamic artifacts, in Gagliardi, P. (Ed.) *Symbols and Artifacts*, pp.147–167, Berlin: De Gruyter.

CLANCEY, W.J., 1993, Situated action: A neuropsychological interpretation response to Vera and Simon, *Cognitive Science*, **17**, 87–116.

CLANCEY, W. J., SMOLIAR, S. W. and STEFIK, M. J. (Eds), 1994, *Contemplating Minds: A Forum for Artificial Intelligence*, Cambridge, MA: MIT Press.

CLARK, H.H. and BRENNAN, S.E., 1991, Grounding in communication, in Levine, J., Resnick, L.B. and Behrend, S.D. (Eds) *Shared Cognition: Thinking as Social Practice*, Washington, DC: APA Books.

CLARK, H.H. and SCHAEFER, E.F., 1989, Contributing to discourse, *Cognitive Science*, **13**, 259–294.

CLARK, H.H. and WILKES-GIBBS, D., 1986, Referring as a collaborative process, *Cognition*, **22**, 1–39.

CLEMENT, A., 1994, Considering privacy in the development of multi-media communications, *Computer-Supported Cooperative Work*, **2** (1–2), 67–88.

COLE, M., 1990, Cultural psychology: A once and future discipline?, in Berman, J.J. (Ed.) *Nebraska Symposium on Motivation: Cross-cultural Perspectives*, pp.279–333, Lincoln: University of Nebraska Press.

COLE, M., 1995, Culture and cognitive development: From cross-cultural research to creating systems of cultural mediation, *Culture and Psychology*, **1** (1), 25–54.

COLLINS, H.M., 1990, *Artificial Experts: Social Knowledge and Intelligent Machines*, Cambridge, MA: MIT Press.

CONTRACTOR, N.S. and EISENBERG, E.M., 1990, Communication networks and new media in organization, in Fulk, J. and Steinfield, C. (Eds) *Organizations and Communication Technology*, pp.145–174, Newbury Park, CA: Sage.

CONTRACTOR, N.S. and SEIBOLD, D.R., 1993, Theoretical frameworks for the study of structuring processes in group decision support systems, *Human Communication Research*, **19** (4), 528–563.

COOPER, G., 1991, Context and its representation, *Interacting with Computers*, **3** (3), 243–252.

CROZIER, M. and FRIEDBERG, E., 1977, *L'Acteur et le système*, Paris: Seuil.

CUSHMAN, P., 1990, Why the self is empty: Toward a historically situated psychology, *American Psychologist*, **45** (5), 599–611.

DAWES, R.M., 1988, *Rational Choice in an Uncertain World*, Orlando, FL: Harcourt Brace Jovanovich.

DEAUX, K., 1992, Personalizing identity and socializing self, in Breakwell, G.M. (Ed.) *Social Psychology of Identity and the Self Concept*, pp.9–34, London: Surrey University Press.

DOBRIN, D., 1989, *Writing and Technique*, Urbana, IL: National Council of Teachers of English.

DOHÉNY-FARINA, S., 1991, *Rhetoric, Innovation, Technology: Case Studies of Technical Communication in Technology Transfers*, Cambridge, MA: MIT Press.

DUBROVSKY, V.J., RIESLER, S. and SETHNA, B.N., 1991, The equalization phenomenon: Status effects in computer-mediated and face-to-face decision-making groups, *Human–Computer Interaction*, **6**, 119–146.

DUNLOP, C. and KLING, R., 1991, Social controversies about computerization, in Dunlop, C. and Kling, R. (Eds) *Computerization and Controversy: Value Conflict and Social Choice*, pp.1–12, Boston, MA: Academic Press.

DURANTI, A. and GOODMAN, C. (Eds), 1992, *Rethinking Context*, Cambridge: Cambridge University Press.

EARNSHAW, R.A., GIGANTE, M.A. and JONES, H., 1993, *Virtual Reality Systems*, London: Academic Press.

ECO, U., 1979, *Lector in fabula*, Milano: Bompiani.

ECO, U., 1994, *Six Walks in the Fictional Woods: Norton Lectures 1993*, Cambridge, MA: Harvard University Press.

EGIDO, C., 1990, Teleconferencing as a technology to support cooperative work: Its possibilities and limitations, in Galegher, J., Kraut, R.E. and Egido, C. (Eds) *Intellectual Teamwork: Social and Technological Foundations of Cooperative Work*, pp.351–372, Hillsdale, NJ: Erlbaum.

EHN, P., 1988, *Work-oriented Design of Computer Artifacts*, Stockholm: Arbetslivscentrum.

ELLIS, C.A., 1991, The socialization of computers, in Stamper, R.K., Kerola, P., Lee, R. and Lyytinen, K. (Eds) *Collaborative Work, Social Communication and Information Systems*, pp.373–385, Amsterdam: North-Holland.

ELLIS, C.A., GIBBS, S.J. and REIN, G.L., 1991a, Groupware: Some issues and experiences, *Communications of the ACM*, **34** (1), 38–58.

ELLIS, C.A. and WAINER, J., 1994, Goal-based models of collaboration, *Collaborative Computing*, **1** (1), 61–86.

ELLIS, S., KAISER, M. and GRUNEWALD, A., 1991b, *Pictorial Communication in Virtual and Real Environments*, London: Taylor & Francis.

EMERY, F.E. and TRIST, E., 1969, *Form and Content in Industrial Democracy*, London: Tavistock.

ERICCSON, K.A. and SIMON, H.A., 1984, *Protocol Analysis: Verbal Reports as Data*, Cambridge, MA: MIT Press.

EVELAND, J.D. and BIKSON, T.E., 1987, Evolving electronic communication networks: An empirical assessment, *Office: Technology and People*, **3**, 103–128.

FAFCHAMPS, D., REYNOLDS, D. and KUCHINSKY, A., 1991, The dynamics of small group decision-making using electronic mail, in Bowes, J.M. and Benford, S.B. (Eds) *Studies in Computer-Supported Cooperative Work*, Amsterdam: North-Holland.

FAYOL, H., 1949, *General and Industrial Management*, London: Pitman.

FELDMAN, M., 1987, Electronic mail and weak ties in organizations, *Office: Technology and People*, **3**, 83–102.

FISHHOFF, B., 1986, Decision making in complex systems, in Hollnagel, E., Mancini, G. and Woods, D.D. (Eds) *Intelligent Decision Support in Process Environments*, pp.61–86, Berlin: Springer.

FITCH, M., HUSTON, A. and WRIGHT, J., 1993, From television form to genre schemata: Children's perceptions of television reality, in Berry, G.L. and Keiko Asamen, J. (Eds) *Children and Television*, pp.38–52, Newbury Park, CA: Sage.

FLORES, F., GRAVES, M., HARTFIELD, B. and WINOGRAD, T., 1988, Computer systems and the design of organizational interaction, *ACM Transactions on Office Information Systems*, **6** (2), 153–172.

FOWLES, J., 1969, *The French Lieutenant's Woman*, New York: Penguin.

FRIJDA, N.H., 1986, *The Emotions*, Cambridge: Cambridge University Press.

FRIJDA, N.H., 1987, Emotions, cognitive structures and action tendency, *Cognition and Emotion*, **1**, 115–143.

FRIJDA, N.H. and SWAGERMAN, J., 1987, Can computers feel? Theory and design of an emotional system, *Cognition and Emotion*, **1**, 235–258.

FULK, J., 1993, Social construction of communication technology, *Academy of Management Journal*, **36** (5), 921–950.

FULK, J., SCHMITZ, J.A. and SCHWARZ, D., 1992, The dynamics of context-behaviour interactions in computer-mediated communications, in Lea, L. (Ed.) *Contexts of Computer-Mediated Communication*, pp.7–29, Hemel Hempstead: Harvester Wheatsheaf.

FULK, J. and STEINFIELD, C. (Eds), 1990, *Organizations and Communication Technology*, Newbury Park, CA: Sage.

GALEGHER, J., KRAUT, R.E. and EGIDO, C. (Eds), 1990, *Intellectual Teamwork: Social and Technological Foundations of Cooperative Work*, Hillsdale, NJ: Erlbaum.

GASSER, L., 1991, Social conceptions of knowledge and action: Distributed Artificial Intelligence foundations and open systems semantics, *Artificial Intelligence*, **47**, 107–138.

GELERNTER, D., 1991, *Mirror Worlds*, New York: Oxford University Press.

GENESERETH, M.R. and KETCHPEL, S.P., 1994, Software agents, *Communications of the ACM*, **37** (7), 48–53.

GERGEN, K.J. and GERGEN, M.M., 1988, Narrative and the self as relationship, in Berkowitz, L. (Ed.) *Advances in Experimental Social Psychology*, pp.17–56, San Diego, CA: Academic Press.

GERGEN, K.J., 1991, *The Saturated Self*, New York: Basic Books.

GERSON, E.M. and STAR, S.L., 1986, Analyzing due processes in the workplace, *ACM Transactions on Office Information Systems*, **4** (3), 257–270.

GIBSON, J.J., 1986, *The Ecological Approach to Visual Perception*, Hillsdale, NJ: Erlbaum (1st edn, 1979).

GIBSON, W., 1984, *Neuromancer*, New York: Ace Books.

GIDDENS, A., 1984, *The Constitution of Society*, Cambridge: Polity Press.

GIDDENS, A., 1991, *Modernity and Self-identity*, Oxford: Polity Press.

GILBERT, D. T., 1991, How mental systems believe, *American Psychologist*, **46** (2), 107–119.

GILBERT, G.N. and Mulkay, M.J., 1984, *Opening Pandora's Box: A Sociological Analysis of Scientific Discourse*, Cambridge: Cambridge University Press.

GLADWIN, T., (1964), Culture and logical process, in W. Goodenough (Ed.) *Explorations in Cultural Anthropology: Essays presented to George Peter Murdock*, pp.156–187, New York: McGraw-Hill.

GLADWIN, T., (1970), *East is a Big Bird*, Cambridge, MA: Harvard University Press.

GOODMAN, P.S. and Sproull, L.S. (Eds), 1990, *Technology and Organizations*, San Francisco, CA: Jossey-Bass.

GOODWIN, C. and HERITAGE, J., 1990, Conversation analysis, *Annual Review of Anthropology*, **19**, 283–307.

GREEN, E., OWEN, J. and PAIN, D., 1991, Office system development and gender: Implications for computer supported cooperative work, in Bannon, L., Robinson, M. and Schmidt, K. (Eds) *Proc. 2nd European Conference on Computer-Supported Cooperative Work*, pp.33–47, Dordrecht: Kluwer.

GREENBERG, S. (Ed.), 1991, *Computer-supported Cooperative Work and Groupware*, London: Academic Press.

GREENO, J.G., 1989, Situations, mental models and generative knowledge, in Klahr, D. and Kotowsky, K. (Eds) *Complex Information Processing: The Impact of H.A. Simon*, pp.285–318, Hillsdale, NJ: Erlbaum.

GREENO, J.G. and MOORE, J. L., 1993, Situativity and symbols: Response to Vera and Simon, *Cognitive Science*, **17**, 49–59.

GREIF, I., 1988, Introduction, in Greif, I. (Ed.) *Computer-supported Cooperative Work: A Book of Readings*, San Mateo, CA: Morgan Kaufman.

GRUDIN, J., 1994, Groupware and social dynamics: Eight challenges for developers, *Communications of the ACM*, **37** (1), 92–105.

GUTEK, B.A., 1990, Work group structure and information technology: A structural contingency approach, in Galegher, J., Kraut, R.E. and Egido, C. (Eds) *Intellectual Teamwork: Social and Technological Foundations of Cooperative Work*, pp.63–78, Hillsdale, NJ: Erlbaum.

HAKKEN, D., 1993, Computing and social change: New technology and workplace transformation, 1980–1990, *Annual Review of Anthronology*, **22**, 107–132.

HAMMOND, N.V., BARNARD, P.J., MORTON, J. LONG, J.B. and CLARK, I.A., 1987, Characterizing user performance in command-driven dialogue, *Behaviour and Information Technology*, **6** (2), 159–205.

HAYES, N.A. and BROADBENT, D.E., 1988, Two modes of learning for interactive tasks, *Cognition*, **28**, 249–276.

HEATH, C. and LUFF, P., 1993, Disembodied conduct: Interactional asymmetries in video-mediated communication, in Button, G. (Ed.) *Technology in Working Order*, pp.35–54, London: Routledge.

HEIM, M., 1991, The erotic ontology of cyberspace, in Benedikt, M. (Ed.) *Cyberspace: First Steps*, pp.59–80, Cambridge, MA: MIT Press.

HESSE, B.W., WERNER, C.M. and ALTMAN, I., 1987, Temporal aspects of computer-mediated communication, *Computers in Human Behavior*, **4**, 147–165.

HIGGINS, E.T., 1990, Personality, social psychology, and person-situated relations: Standards of knowledge activation as a common language, in Pervin, L.A. (Ed.) *Handbook of Personality*, pp.301–338, New York: Guilford.

HILTZ, S.R., 1988, Productivity enhancement from computer-mediated-communication: A systems contingency approach, *Communications of the ACM*, **31**, 1438–1454.

HILTZ, S.R. and TUROFF, M., 1978, *The Network Nation: Human Communication via Computer*, Reading, MA: Addison-Wesley.

HOGG, M.A. and ABRAMS, D., 1988, *Social Identifications: A Social Psychology of Intergroup Relations and Group Processes*, London: Routledge.

HOGG, M.A. and McGARTY, C., 1990, Self-categorization and social identity, in Abrams, D. and Hogg, M.A. (Eds) *Social Identity Theory*, pp.10–27, Hemel Hempstead: Harvester Wheatsheaf.

HOLLINGSHEAD, A.B., McGRATH, J.E. and O'CONNOR, K.M., 1993, Group task performance and communication technology, *Small Group Research*, **24** (3), 307–333.

HOLYOAK, K.J. and THAGARD, P., 1995, *Mental Leaps – Analogy in Creative Thought*, Cambridge, MA: The MIT Press.

HUBER, G.P., VALACICH, J. S. and JESSUP, L.M., 1993, A theory of the effects of group support systems on an organization's nature and decisions, in Jessup, L.M. and Valacich, J.S. (Eds) *Group Support Systems: New Perspectives*, pp. 78–96, New York: Academic Press.

HURRELMANN, R., 1988, *Social Structure and Personality Development*, Cambridge: Cambridge University Press.

JIROTKA, M., GILBERT, N. and LUFF, P., 1992, On the social organization of organizations, *Computer-Supported Cooperative Work*, **1** (1–2), 95–118.

JOHNSON, M., 1993, *Moral Imagination: Implications of Cognitive Science for Ethics*, Chicago, IL: University of Chicago Press.

JOHNSON-LAIRD, P.N., 1983, *Mental Models*, Cambridge: Cambridge University Press.

JOHNSON-LAIRD, P.N., 1988, *The Computer and the Mind*, Cambridge, MA: Cambridge University Press.

JOHNSON-LAIRD, P.N., 1993, *Human and Machine Thinking*, Hillsdale, NJ: Erlbaum.

JOHNSON-LAIRD, P.N. and Byrne, R.M.J., 1993, Précis of 'Deduction', *Behavioral and Brain Sciences*, **16** (2), 323–333.

JONES, S.G., 1995, Understanding community in the information age, in Jones, S.G. (Ed) *Cybersociety: Computer-Mediated Communication and Community*, pp.10–35, Thousand Oaks, CA: Sage.

KENNEDY, R.S., LANE, N. E., LILIENTHAL, M.G., BERBAUM, K.S. and HETTINGER, L.J., 1992, Profile analysis of simulator sickness symptoms: Application to virtual environment systems, *Presence: Teleoperators and Virtual Environment*, **1** (3), 295–301.

KERKNODE, F., 1979, *The Genesis of Secrecy: On the Interpretation of Narrative*, Cambridge: Cambridge University Press.

KLEIN, H.K. and KRAFT, P., 1994, Social control and social contract in NetWORKing, *Computer-Supported Cooperative Work*, **2** (1–2), 89–108.

KLING, R., 1980, Social analyses of computing: Theoretical perspectives in recent empirical research, *Computing Surveys*, **12**, 61–110.

KLING, R., 1994, Reading 'all about' computerization: How genre conventions shape nonfiction social analysis, *The Information Society*, **10**, 147–172

KLING, R., 1995, *Computerization and Controversy*, San Diego, CA: Academic Press.

KLING, R. and SCACCHI, W., 1982, The web of computing: computer technology as social organization, *Advances in Computers*, **21**, 2–60.

KOTTERMANN, J.E., DAVIS, F.D. and REMUS, W.E., 1994, Computer-assisted decision making: Performance, beliefs, and the illusion of control, *Organizational Behaviour and Buman Decision Processes*, **57**, 26–37.

KRAUT, R.E. and STREETER, L.A., 1995, Coordination in software development, *Communications of the ACM*, **38** (3), 69–81.

KRAUT, R.E., EGIDO, C. and GALEGHER, J., 1990, Patterns of contact and communication in scientific research collaboration, in Galegher, J., Kraut, R.E. and Egido, C. (Eds) *Intellectual Teamwork: Social and Technological Foundations of Cooperative Work*, pp.149–172, Hillsdale, NJ: Erlbaum.

KRAUT, R.E., COOL, C., RICE, R.E. and FISH, R.S., 1994, Life and death of new technology: Task, utility and social influences on the use of a common medium, in Furuta, R. and Neuwirth, C. (Eds) *Proc. Conference on CSCW* (October 1994, Chapel Hill, NC), pp.13–21, New York: ACM Press.

KRUEGER, M.W., 1990, Videoplace and the interface of the future, in Laurel, B. (Ed.) *The Art of Human–Computer Interface*, pp.417–422, Reading, MA: Addison-Wesley.

KRUEGER, M.W., 1991, *Artificial Reality*, Reading, MA: Addison-Wesley.

KUBEY, R. and CSIKSZENTMIHALYI, M., 1990, *Television and the Quality of Life: How Viewing Shapes Everyday Experience*, Hillsdale, NJ: Erlbaum.

LAKOFF, G., 1987, *Women, Fire, and Dangerous Things: What Categories Reveal About the Mind*, Chicago, IL: Chicago University Press.

LANIER, J. and BIOCCA, F., 1992, An insider's view of the future of virtual reality, *Journal of Communication*, **42** (4), 150–172.

LATOUR, B., 1987, *Science in Action*, Milton Keynes: Open University Press.

LAUREL, B., 1991, *Computers as Theater*, Menlo Park, CA: Addison-Wesley.

LAVE, J., 1988, *Cognition in Practice*, Cambridge: Cambridge University Press.

LAVE, J., 1992, Word problems: A microcosm of theories of learning, in Light, P.H., Butterworth, G.E. (Eds) *Contexts and Cognition: Ways of Learning and Knowing*, pp.74–92, Hemel Hempstead: Harvester Wheatsheaf.

LAZARUS, R.S., 1991, *Emotion and Adaptation*, New York: Oxford University Press.

LEA, M., 1991, Rationalist assumptions in cross-media comparisons of computer-mediated communication, *Behaviour and Information Technology*, **10** (2), 153–172.

LEA, M. , O'SHEA, T., FUNG, P. and SPEARS, R., 1992, 'Flaming' in computer-mediated communication, in Lea, M. (Ed.) *Contexts of Computer-mediated Communication*, pp.89–112, Hemel Hempstead: Harvester Weatsheaf.

LEA, M. AND SPEARS, R., 1991, Computer-mediated communication, de-individuation and group decision-making, *International Journal of Man–Machine Studies*, **34**, 283–301.

LEGRENZI, P., GIROTTO, V. and JOHNSON-LAIRD, P.N., 1993, Focusing in reasoning and decision-making, *Cognition*, **49** (1–2), 37–66.

LYNCH, M., 1985, *Art and Artifact in Laboratory Science*, London: Routledge & Kegan Paul.

MCCARTHY, J.C. and MONK, A.F., 1994a, Measuring the quality of computer-mediated communication, *Behaviour and Information Technology*, **13** (5), 311–319.

MCCARTHY, J.C. and MONK, A.F., 1994b, Channels, conversation, cooperation and relevance: All you wanted to know about communication but were afraid to ask, *Collaborative Computing*, **1** (1), 35–60.

MCCAULEY, M.E. and SHARKEY, T.J., 1992, Cybersickness: Perception of self-motion in virtual environments, *Presence*: *Teleoperators and Virtual Environments*, **1** (3), 311–318.

MCGRATH, J.E., 1990, Time matters in groups, in Galegher, J., Kraut, R.E. and Egido, C. (Eds) *Intellectual Teamwork*: *Social and Technological Foundations of Cooperative Work*, pp.23–61, Hillsdale, NJ: Erlbaum.

MCGRATH, J.E. and HOLLINGSHEAD, A.B., 1993, Putting the 'Group' back in group support systems: Some theoretical issues about dynamic processes in groups with technological enhancements, in Jessup, L.M. and Valacich, J.S. (Eds) *Group Support Systems*: *New Perspectives*, pp.78–96, New York: Academic Press.

MCGRATH, J.E. and HOLLINGSHEAD, A.B., 1994, *Groups Interacting with Technology*, London: Sage.

MCGRATH, J.E. and KELLY, J.R., 1986, *Time and Human Interaction*: *Towards a Social Psychology of Time*, New York: Guilford.

MCGUIRE, W.J., 1983, A contextualist theory of knowledge: Its implications for innovation and reform in psychological research, in Berkowitz, L. (Ed.) *Advances in Experimental Social Psychology*, pp.1–47, New York: Academic Press.

MCGUIRE, W.J. and MCGUIRE, C.V., 1988, Content and process in the experience of self, in Berkowitz, L. (Ed.) *Advances in Experimental Social Psychology*, pp. 97–144, San Diego, CA: Academic Press.

MCLEOD, P.L., 1992, An assessment of the experimental literature on electronic support of group work: Results of a meta-analysis, *Human–Computer Interaction*, **7**, 257–280.

MCQUAIL, D., 1994, *Mass Communication Theory* (3rd edn), London: Sage.

MAES, P., 1994, Agents that reduce work and information overload, *Communications of the ACM*, **37** (7), 30–40.

MALONE, T. and ROCKART, J.F., 1991, Computers, networks and the corporation, *Scientific American*, **265** (3), 92–99.

MANSELL, R., 1993, *The New Telecommunications*, London: Sage.

MANTOVANI, G., 1991 *La qualità dell'interazione uomo–computer*, Bologna: Il Mulino.

MANTOVANI, G., 1994a, *Was der Computer mit uns macht*: *Sozialpsychologische Aspekte der Communication mit und durch den Computer*, Mainz: Matthias-Grunewald Verlag.

MANTOVANI, G., 1994b, Is computer-mediated communication intrinsically apt to enhance democracy in organizations?, *Human Relations*, **47** (1), 45–62.

MANTOVANI, G., 1995, Virtual reality as a communication environment: Consensual hallucination, fiction and possible selves, *Human Relations*, **48**, (1), 669–683.

MANTOVANI, G., 1996, Social context in HCI: A new framework for mental models, cooperation and communication, *Cognitive Science* (in press).

MANTOVANI, G. and BOLZONI, M., 1994, Analysing and evaluating multi-actor multi-goal systems in use: Social contexts and participation in three Vocational Guidance Systems, *Behaviour and Information Technology*, **13** (3), 201–214.

MARCH, J.G., 1991, How decisions happen in organizations, *Human–Computer Interaction*, **6**, 95–117.

MARCH, J.G. and OLSEN, J.P., 1989, *Rediscovering Institutions*, New York: Free Press.

MARKUS, H. and CROSS, S., 1990, The interpersonal self, in Pervin, L.A. (Ed.) *Handbook of Personality*, pp.576–608, New York: Guilford.

MARKUS, H. and KUNDA, Z., 1986, Stability and malleability of the self-concept, *Journal of Personality and Social Psychology*, **51**, 858–866.

MARKUS, H. and NURIUS, P., 1986, Possible selves, *American Psychologist*, **41** (9), 954–969.

MARKUS, H. and WULF, E., 1987, The dynamic self-concept: A social psychological perspective, *Annual Review of Psychology*, **38**, 299–337.

MARSHAK, R.T., 1993, Action technologies' workflow products, *Workgroup Computing Report*, **16** (5), 1–24.

MEAD, G.H., 1934, *Mind, Self and Society*, Chicago: University of Chicago Press.

MEDINA-MORA, R., WINOGRAD, T., FLORES, R. and FLORES, F., 1992, The action workflow approach to workflow management technology, *Proc. CSCW 1992*, pp.281–288, New York: ACM.

MEINONG, A., 1904, On the theory of objects, in Chisholm, R.M. (Ed.) *Realism and the Background of Phenomenology*, (1960), New York: Free Press.

MILLER, J.R., Galanter, E. and Pribram, K., 1960, *Plans and the Structure of Behavior*, New York: Holt, Rinehart & Winston.

MINSKY, M. and RIECKEN, D., 1994, A conversation with Marvin Minsky about agents, *Communications of the ACM*, **37** (7), 22–29.

MITCHELL, T.R. and BEACH, L.R., 1990, '... Do I love thee? Let me count...'. Towards an understanding of intuitive and automatic decision making, *Organizational Behaviour and Human Decision Processes*, **47**, 1–20.

NISBETT, R.E. and WILSON, T.D., 1977, Telling more than we know: Verbal reports on mental processes, *Psychological Review*, **84** (3), 231–259.

NISSENBAUM, H., 1994, Computing and accountability, *Communications of the ACM*, **37** (1), 72–80.

NORMAN, D.A., 1988, *The Psychology of Everyday Things*, New York: Basic Books.

NORMAN, D.A., 1990, Why interfaces don't work, in Laurel, B. (Ed.) *The Art of Human–Computer Interface Design*, pp.209–219, Reading, MA: Addison-Wesley.

NORMAN, D.A., 1991, *Cognitive Artifacts*, in Carroll, J.M. (Ed.) *Designing Interface: Psychology at the Human–Computer Interface*, pp.17–39, Cambridge: Cambridge University Press.

NORMAN, D.A., 1992, *Turn Signals are the Facial Expressions of Automobiles*, Reading, MA: Addison-Wesley.

NORMAN, D.A., 1993, *Things that Make us Smart*, Reading, MA: Addison-Wesley.

NORMAN, D.A., 1994, How might people interact with agents, *Communications of the ACM*, **37** (7), 68–71.

OLSON, M.H. and BLY, S.A., 1991, The Portland Experience: A report on a distributed research group, *International Journal of Man–Machine Studies*, **34**, 211–228.

PAVEL, T.G., 1986, *Fictional Worlds*, Cambridge, MA: Harvard University Press.

PAYNE, J.W., Bettman, J.R. and Johnson, E.J., 1992, Behavioral decision research: A constructive processing perspective, *Annual Review of Psychology*, **43**, 87–131.

PERROW, C., 1970, *Organizational Analysis: A Sociological View*, Belmont, CA: Wadsworth.

PERSE, E.M. and Courtright, J.A., 1993, Normative images of communication media, *Human Communication Research*, **19** (4), 485–503.

PINCH, T.J. and Bijker, W.E., 1993, The social construction of facts and artifacts, in Bijker, W.E., Hughes, T.P. and Pinch, T.J. (Eds) *The Social Construction of Technological Systems*, pp.17–50, Cambridge, MA: MIT Press.

POOLE, M.S. and DeSANCTIS, G., 1990, Understanding the use of group decision support systems: The theory of adaptive structuration, in Fulk, J. and Steinfield, C. (Eds) *Organizations and Communication Technology*, pp.173–193, Newbury Park, CA: Sage.

POOLE, M.S. and DESANCTIS, G., 1992, Microlevel structuration in computer-supported group decision making, *Human Communication Research*, **19** (1), 5–49.

POPPER, K.R. and ECCLES, J.C., 1977, The Self and its Brain, Berlin: Springer.

RASMUSSEN, J., BREHMER, B. and LEPLAT, J., 1991, *Distributed Decision Making: Cognitive Models for Cooperative Work*, Chichester: Wiley.

REID, E., 1995, Virtual worlds: Culture and imagination, in Jones, S.G. (Ed) *Cybersociety: Computer-mediated Communication and Community*, pp.164–183, Thousand Oaks, CA: Sage.

RESNIK, L.B., 1994, Situated rationalism: Biological and social preparation for learning, in Hirschfeld, L.A. and Gelman, S.A. (Eds) *Mapping the Mind: Domain Specificity in Cognition and Culture*, pp.474–493, Cambridge: Cambridge University Press.

RHEINGOLD, H.R., 1991, *Virtual Reality*, New York: Summit Books.

RHEINGOLD, H.R., 1993, A slice of life in my virtual community, in Harasim, L.M. (Ed) *Global Networks*, pp.57–80, Cambridge, MA: MIT Press.

RICE, R.R., 1986, Applying the human relations perspective to the study of new media, *Computer and Society*, **15**, 32–37.

RICE, R.R., 1990, Computer-mediated communication system network data: Theoretical concerns and empirical examples, *International Journal of Man–Machine Studies*, **32**, 627–647.

RICE, R.R., 1993, Media appropriateness: Using social presence theory to compare traditional and new organizational media, *Human Communication Research*, **19** (4), 451–484.

RICE, R.R. and LOVE, G., 1987, Electronic emotion: A content and network analysis of a computer-mediated communication network, *Communication Sciences*, **14**, 85–105.

RICE, R.R. and SHOOK, D., 1990, Voice messaging, coordination and communication, in Galegher, J., Kraut, R. and Egido, R. (Eds) *Intellectual Teamwork: Social and Technological Foundations of Co-operative Work*, pp.327–350, Hillsdale, NJ: Erlbaum.

ROBINSON, M., 1991, Double-level languages and co-operative working, *AI & Society*, **5**, 34–60.

ROGOFF, B., BAKER-SENNETT, J. and MATUSOV, E., 1994, Considering the concept of planning, in Haith, M.M., Benson, J.B., Roberts, R.J. and Pennington, B.F. (Eds) *The Development of Future Oriented Processes*, pp.353–373, Chicago: Chicago University Press.

ROGOFF, B., CHAVAJAY, P. and MATUSOV, E., 1993, Questioning assumptions about culture and individuals, *Behavioral and Brain Sciences*, **16**, 533–534.

ROSALDO, M., 1984, Towards an anthropology of self and feeling, in Schroeder, R. and Le Vine, R. (Eds) *Culture Theory: Essays on Mind, Self and Emotion*, pp.137–158, Cambridge: Cambridge University Press.

ROSENBERG, M., 1981, The self-concept: Social production and social force, in Rosenberg, M. and Turner, R.H. (Eds) *Social Psychology: Sociological Perspectives*, pp.593–624, New York: Basic Books.

ROSS, L. and NISBETT, R.E., 1991, *The Person and the Situation*, New York: McGraw-Hill.

ROUSSEAU, D.M., 1983, Technology in organizations: A constructive review and analytic framework, in Seashore, S.E., Lawler, E.E., Mirvis, P.H. and Camman, C. (Eds) *Assessing Organizational Changes*, New York: Wiley.

SAHLINS, M., 1985, *Islands of History*, Chicago, IL: University of Chicago Press.

SAUNDERS, C.S., ROBEY, D. and VAVEREK, K.A., 1994, The persistence of status differentials in computer conferencing, *Human Communication Research*, **20** (4), 443–472.

SCHMIDT, K., 1991, Riding a tiger, or Computer-Supported Cooperative Work, in Bannon, L., Robinson, M. and Schmidt, K. (Eds) *E-CSCW'91*, Dordrecht: Kluwer.

SCHMIDT, K., 1994, The organization of cooperative work: Beyond the 'Leviathan' conception of the organization of cooperative work, in Furuta, R. and Neuwirth, C. (Eds) *Proc. Conf. on CSCW* (October 1994, Chapel Hill, NC), pp.101–112, New York: ACM Press.

SCHMIDT, K. and BANNON, L., 1992, Taking CSCW seriously, *Computer-Supported Cooperative Work*, **1**, 7–40.

SCHNEIDER, K. and WAGNER, I., 1993, Constructing the 'Dossier Représentatif'. Computer-based information-sharing in French hospitals, *Computer-Supported Cooperative Work*, **1** (4), 229–253.

SCHNEIDERMAN, B., 1992, *Designing the User Interface: Strategies for Effective Human–Computer Interaction*, Reading, MA: Addison-Wesley.

SCHULER, D., 1994, Community networks: Building a new participatory medium, *Communications of the ACM*, **37** (1), 38–51.

SHAPIRO, M.A. and MCDONALD, D.G., 1992, Virtual reality: Implications of virtual reality for judgments about reality, *Journal of Communication*, **42** (4), 94–114.

SHIPPEY, T., 1993, Inside the screen, *The Times Literary Supplement*, April 30.

SHWEDER, R.A. and SULLIVAN, M.A., 1993, Cultural psychology: Who needs it?, *Annual Review of Psychology*, **44**, 497–523.

SIEGEL, J., DUBROVSKY, V., KIESLER, S. and MCGUIRE, T.W., 1986, Group processes in computer-mediated communication, *Organizational Behaviour and Human Decision Processes*, **37**, 157–187.

SIMON, H.A., 1960, *The New Science of Management Decision*, New York: Harper & Row.

SIMON, H.A., 1981, *The Sciences of the Artificial*, Cambridge, MA: MIT Press.

SIMON, H.A., 1983, *Reason in Human Affairs*, Oxford: Blackwell.

SMITH, J., 1982, The bare facts of ritual, in *Imagining Religion: From Babylon to Jamestown*, pp.53–65, Chicago: Chicago University Press.

SMITH, N., BIZOT, E. and HILL, T., 1988, *Use of Electronic Mail in a Research and Development Organization*, Tulsa, OK: University of Tulsa.

SPEARS, R. and LEA, M., 1992, Social influence and the influence of the "social" in computer-mediated communication, in Lea, L. (Ed.) *Contexts of Computer-mediated Communication*, pp.30–65, Hemel Hempstead: Harvester Wheatsheaf.

SPEARS, L., LEA, M. and LEE, S., 1990, De-individuation and group polarization in computer-mediated communication, *British Journal of Social Psychology*, **29**, 121–134.

SPELLER, G.J. and BRANDON, J.A., 1986, Ethical dilemmas constraining the use of expert systems, *Behaviour and Information Technology*, **5**, 141–143.

SPITULNIK, D., 1993, Anthropology of mass media, *Annual Review of Anthropology*, **22**, 293–315.

SPROULL, L. and KIESLER, S., 1991, *Connections: New Ways of Working in the Networked Organizations*, Cambridge, MA: MIT Press.

STAMPER, R.K., LEE, R., KEROLA, P. and LYYTINEN, K. (Eds), 1991, *Collaborative Work, Social Communications and Information Systems*, Amsterdam: North-Holland.

STASSER, G., 1992, Pooling of unshared information during group discussion, in Worchell, S., Wood, W. and Simpson, J.A. (Eds) *Group Processes and Productivity*, pp.48–67, Newbury Park, CA: Sage.

STEUER, J., 1992, Defining virtual reality: Dimensions determining telepresence, *Journal of Communication*, **42** (4), 73–93.

STINCHOMBE, A.L., 1990, *Information and Organizations*, Berkeley, CA: University of California Press.

STONE, A.R., 1991, Will the real body please stand up?: Boundary stories about virtual cultures, in Benedikt, M. (Ed.) *Cyberspace: First Steps*, pp.81–118, Cambridge, MA: MIT Press.

STONE, A.R., 1992, *Cyberspace II*, Cambridge, MA: MIT Press.

SUCHMAN, L., 1987, *Plans and Situated Actions*, Cambridge: Cambridge University Press.

SUCHMAN, L., 1993, Do categories have politics? The language/action perspective reconsidered, in De Michelis, G., De Simone, C. and Schmidt, K. (Eds) *Proc. Third European Conference on Computer-Supported Cooperative Work*, pp.1–14, Dordrecht: Kluwer.

SUCHMAN, L.A. and TRIGG, R.H., 1991, Understanding practice: Video as a medium for reflexion and design, in Greenbaum, J. and Kyng, M. (Eds) *Design at work: Cooperative Design of Computer Systems*, pp.65–90, Hillsdale, NJ: Erlbaum.

TAINSH, M.A., 1988, The concept of an information management system and its use within design studies, *Behaviour and Information Technology*, **7** (4), 443–455.

TAJFEL, H., 1972, Experiments in a vacuum, in Israel, J. and Tajfel, H. (Eds) *The Context of Social Psychology: A Critical Assessment*, pp.19–38, London: Academic Press.

TAJFEL, H. and TURNER, J. C., 1986, The social identity theory of intergroup behaviour, in Worchell, S. and Austin, W.G. (Eds) *Psychology of Intergroup Relations*, pp.7–24, Chicago: Nelson-Hall.

TATAR, D.G., FOSTER, G. and BOBROW, D.G., 1991, Design for conversation: Lessons from Cognoter, *International Journal of Man–Machine Studies*, **34**, 185–209.

TAUBER, M.J. and ACKERMANN, D. (Eds), 1990, *Mental Models and Human–Computer Interaction*, Vol. 2, Amsterdam: North-Holland.

TAYLOR, C., 1989, *Sources of the Self*, Cambridge, MA: Harvard University Press.

TAYLOR, C., 1991, *The Ethics of Authenticity*, Cambridge MA: Harvard University Press.

TAYLOR, F.W., 1911, *Principles of Scientific Management*, New York: Harper & Row.

TREVINO, L.K., DAFT, R.L. and LENGEL, R.H., 1990, Understanding managers' media choices: A symbolic interactionist perspective, in Fulk, J. and Steinfield, C. (Eds) *Organizations and Communication Technology*, pp.71–94, Newbury Park, CA: Sage.

TRIST, E.L., 1981, The sociotechnical perspective, in van der Ven, A.H. and Joyce, W.F. (Eds) *Perspectives on Organization, Design and Behavior*, New York: Wiley.

TURNER, J., HOGG, M.A., OAKES, P.J., REICHER, S.D. and WETHERELL, M.S., 1987, *Rediscovering the Social Group: A Self-categorization Theory*, Oxford: Blackwell.

VALACICH, J.S., DENNIS, A.R. and CONNOLLY, T., 1994, Idea generation in computer-based groups: A new ending to an old story, *Organizational Behaviour and Human Decision Processes*, **57**, 448–467.

VAN GELDER, L., 1991, The strange case of the electronic lover, in Dunlop, C. and Kling, R. (Eds) *Computerization and Controversy*, pp. 364–375, Boston, MA: Academic Press.

VEER, VAN DER, G.C., GUEST, S., HASEKAGER, P., INNOCENT, P., MCDAID, L., OSTREICHER, L., TAUBER, M.J., VOS, U., WAERN, I., 1990, Designing for the mental model: An interdisciplinary approach to the definition of a user interface for electronic mail systems, in Ackerman, D. and Tauber, M.J. (Eds) *Mental Models and Human–Computer Interaction*, Vol. 1, pp.253–288, Amsterdam: North-Holland.

VERA, A.H. and SIMON, H.A., 1993, Situated action: A symbolic interpretation, *Cognitive Science*, **17**, 7–48.

WALTHER, J.B., 1994, Anticipated ongoing interaction versus channel effects on relational communication in computer-mediated interaction, *Human Communication Research*, **20** (4), 473–501.

WALTHER, J.B. and BURGOON, J.K., 1992, Relational communication in computer-mediated interaction, *Human Communication Research*, **19** (1), 50–88.

WALTON, K., 1984, Do we need fictional entities? Notes toward a theory, in AA.VV., *Aesthetics: Proc. 8th Int. Wittgenstein Symposium*, Vienna: Hodler–Pichler–Temsky.

WARNER, R.M., 1988, Rhythm in social interaction, in McGrath, J.E. (Ed.) *The Social Psychology of Time: New Perspectives*, pp.63–88, Newbury Park, CA: Sage.

WEEDMAN, J., 1991, Task and non-task functions of a computer conference used in professional education: A measure of flexibility, *International Journal of Man–Machine Studies*, **34**, 303–318.

WEICK, K.E., 1990, Technology as equivoque: Sensemaking in new technologies, in Goodman, P.S. and Sproull, L.S. (Eds) *Technology and Organizations*, San Francisco, CA: Jossey-Bass.

WEISBAND, S., 1994, Overcoming social awareness in computer-supported groups: Does anonymity really help?, *Computer-Supported Cooperative Work*, **2** (4), 285–297.

WEISER, M., 1991, The computer for the 21st century, *Scientific American*, **265** (3), 66–75.

WEXELBLAT, A. (Ed.), 1993, *Virtual Reality: Applications and Explorations*, Boston: Academic Press.

WHITTAKER, S., GEELHOED, E. and ROBINSON, E., 1993, Shared workplaces: How do they work and when are they useful?, *International Journal of Man–Machine Studies*, **39**, 813–842.

WILLIAMS, F. and GIBSON, D.V., 1990, *Technology Transfer: A Communication Perspective*, Newbury Park, CA: Sage.

WILLIAMSON, O.E., 1986, *Economic Organization*, Hemel Hempstead: Harvester Wheatsheaf.

WILSON, P., 1991, *Computer-Supported Cooperative Work*, Oxford: Intellect.

WINNER, L., 1980, Do artifacts have politics?, *Daedalus*, **109**, 121–136.

WINOGRAD, T., 1994, Categories, disciplines and social coordination, *Computer-Supported Cooperative Work*, **2**, 191–197.

WINOGRAD, T. and FLORES, S., 1986, *Understanding Computers and Cognition*, Norwood, NJ: Ablex.

WOODS, D.D. and ROTH, E.M., 1988, Cognitive engineering: Human problem-solving with tools, *Human Factors*, **30** (4), 415–430.

WOODWARD, J., 1965, *Industrial Organization: Theory and Practice*, London: Oxford University Press.

WOOLLEY, B., 1992, *Virtual Worlds*, Oxford: Blackwell.

WYER, R.S. and SRULL, T., 1983, *Handbook of Social Cognition*, Hillsdale, NJ: Erlbaum.

ZALESNY, M.D. and FORD, J.K., 1990, Extending the social information processing perspective: New links to attitudes, behaviours and perceptions, *Organizational Behaviour and Human Decision Processes*, **47**, 205–246.

ZHANG, J. and NORMAN, D.A., 1994, Representations in distributed cognitive tasks, *Cognitive Science*, **18**, 87–122.

ZMUD, R.W., 1990, Opportunities for strategic information manipulation through new information technology, in Fulk, J. and Steinfield, C. (Eds) *Organizations and Communication Technology*, pp.95–116, Newbury Park, CA: Sage.

ZUBOFF, S., 1988, *In the Age of the Smart Machine*, New York: Basic Books.

Index